The Relationship Between Cues from Loop Quantum Gravity and Superradiation

The Relationship Between Cues from Loop Quantum Gravity and Superradiation

Wen - Xiang Chen

ELIVA PRESS

Published by Eliva Press
Email: info@elivapress.com
Website: www.elivapress.com

ISBN: 978-1-63648-373-3

© Eliva Press, 2021
© Wen - Xiang Chen
Cover Design: Eliva Press
Cover Image: Freepik Premium
Printed at: see last page

Wen-Xiang Chen[1][2]

September 30, 2021

[1]University of Department of Astronomy, School of Physics and Materials Science, GuangZhou University
[2]wxchen4277@qq.com

Contents

Chapter 1

The limit y of the incident particle under the superradiance and the Hawking radiation of Kerr black hole

Abstract: It is important to emphasize that as the reference has proved well, the critical storage line of black hole field system is universal, that is, different scalar field coupling functions with the same weak field behavior are $\{C(\phi)\}, C(\phi) = 1 + \alpha\phi^2 + o(\phi^4)$, and have the same function behavior $\alpha = \alpha(\mu); a/M$. The purpose of this paper is to find out the limit y of the incident particle under the superradiance of the preset boundary ($\mu = y\omega$) and get the limit y of the incident particle under the Hawking radiation of the preset boundary ($y\mu = \omega$).

Keywords: The limit y, superradiance, Hawking radiation

1.1 Introduction

In 1972, press and teukolsky cited press 1973 perturbation to propose that it is possible to add a mirror outside the black hole to make a black hole bomb (according to the current explanation, it is a scattering process involving classical mechanics and quantum mechanics.

When a boson wave impacts a rotating black hole, if the wave frequency is located below the superradiance region, the wave reflected from the event horizon will be amplified

$$0 < \omega < m\Omega_H, \Omega_H = \frac{a}{r_+^2 + a^2}, \tag{1.1}$$

where m is azimuthal number of the bosonic wave mode, Ω_H is the angular velocity of black hole horizon. This amplification is superradiant scattering. This amplification is called superradiance scattering. Therefore, the rotational energy of the black hole can be extracted by the superradiance process. If there is a mirror between the horizon of the black hole and the infinite space, the amplified wave will scatter back and forth and grow exponentially, which will cause the SUPERRADIATION of the black hole to become unstable.

The paper presents a normal no-hair theorem [1–3] and reveals an interesting physical fact that in the compound Einstein scalar theory, the outer region of the asymptotically flat black hole with the space regular horizon cannot support static scalar field configuration. This physically interesting property also characterizes black hole spacetime, in which scalar field is non minimum coupled with Ricci curvature scalar of spacetime.

The existence of the non-linear dispersion relationship makes it difficult to define the energy region and horizon. This problem is related to the so-called trans-Planckian problem of Hawking radiation. In Hawking's original derivation, it can be seen that if people track the infinite (observed) radiation emitted from the future, this will have an infinite blue shift. Therefore, Hawking's results seem to depend on the validity of quantum field theory in arbitrarily high-energy curved space and time, and there is no real proof in this regard. In fact, the microphysics of time and space beyond the Planck scale may be very different and contain descriptions that deviate from the general theory of relativity. One of the main interests of gravitational analogues is to provide a framework to study possible deviations from the Lorentz invariant physics of curved space and time. In particular, a simulation model refers to a low-energy geometric description of a system that emerges from a well-known microscopic background. One possible way for microphysics to work is to modify the dispersion relationship of fluctuations at high (er) frequencies, as we have seen in Boglubov dispersion. In this case, the healing length and the Planck length play a similar role. Hawking radiation with this modification has been extensively studied, and it has been proven to have a strong effect on the dispersion effect and maintain its thermal properties under appropriate conditions. In a similar black hole, it is instructive to study the scattering of Hawking radiation fluctuating on the acoustic line of sight, which is a very different mechanism from the case of gravity. In the special case of Bose-Einstein condensation, studies have shown that even in configurations far beyond the hydrodynamic limit, the microscopic mechanism of pair generation is conceptually the same.

Interestingly, [4–9] this has been recently confirmed in the interesting physical works in the space-time of spherical symmetric asymptotically flat rotating black hole, the space rule no mass scalar field configuration can be supported, characterized by the non minimum coupling between the form and the scalar field after the preset boundary. This phenomenon, called spontaneous scaling of black holes, has been studied in the references. In the physics interesting large mass scalar field configuration, it is not least coupled with the space-time of rotating black holes.

If the coupling function of nontrivial scalar field is characterized by weak field functional behavior $C(\varphi) = 1 + \alpha\varphi^2 + o(\varphi^4)$, then the space-time of Kerr black hole is an effective solution to the field equation under the limit of ordinary $\varphi \equiv 0$ limit. This is a physical ideal property of the non ordinary coupled Einstein scalar theory. Here, $\alpha > 0$ is a dimensionless physical parameter, which determines the non minimum coupling strength between scalar field and preset boundary conditions in the space-time of a rotating black hole.

The numerical results given in the interesting work show that for the given value of the non dimensional rotation mass ratio of black hole space-time to a/m [10] the non trivial coupling Einstein scalar system has critical memory online $\alpha = \alpha(\mu)$; It marks the boundary space-time between the non least coupled large mass scalar field configuration of a multi hairy Kerr black hole and a bald (scalar free) Kerr black hole (where μ is the appropriate mass of the non least coupled scalar field). In particular, the critical memory line of the composite system corresponds to the space regularized linearized field configuration black hole supported by the central Kerr (the term "scalar cloud" is usually used in the physical literature [11, 12] to describe these linearized scalar field configurations

6

on the critical existence line of the system).

It is important to emphasize that as the reference has proved well, the critical storage line of black hole field system is universal, that is, different scalar field coupling functions with the same weak field behavior are $\{C(\varphi)\}$, $C(\varphi) = 1 + \alpha\varphi^2 + o(\varphi^4)$, and have the same function behavior $\alpha = \alpha(\mu); a/M)$.

In a recent article [19], it proposes to use the information flow from the simulation model to the gravity situation, which is to understand the superradiation phenomenon from the Bose-Einstein condensation. We have already said that the opposite arrow of this analogy can also be interesting. Here, we will use this point of view to reconsider the stability of quantized vortices in BECs, and study the special instabilities that occur in non-uniform flow BECs similar to hydrodynamic parallel shear flows.

We show how the geometric description of sound propagation measured by relativistic acoustics emerges from these excited Pogripov linear problems. For non-uniformly moving aggregates with supersonic flow regions, this acoustic metric can have the shape of the black hole metric of general relativity. Even if the solutions of Einstein's equations generally cannot be reproduced in these fluid analogs, interesting toy models of acoustic space and time can be obtained, sharing the main interesting characteristics of black holes, such as the horizon and energy regions. This gravitational analogy allows the study of field theory in curved space and time, thanks to the low temperature of the Bose-Einstein condensate in terms of its quantum. We explained in detail how the Pogriupov excitation field is related to the Klein-Gordon field, and discussed that the difference between the two may not be a mistake, but a feature of the gravitational analogue. As a guiding example, we reviewed some studies of Hawking radiation in a simple similar black hole. In the discussion, we introduced two different concepts of instability, which will be of great significance in this article: energy and dynamic instability. Energy instability corresponds to the existence of negative energy modes, while dynamical instability refers to the existence of exponentially increasing zero energy modes. Through the black hole laser, we have seen the first example of the interaction between these two concepts, and we will see it play a role many times throughout our work. We see that energy instability is the basis of Hawking emission in Bose-Einstein condensate.

The purpose of this paper is to find out the limit y of the incident particle under the superradiance of the preset boundary ($\mu = y\omega$) and the limit y of the incident particle under the Hawking radiation of the preset boundary ($y\mu = \omega$).

1.2   Description of the system

We will study the physical and mathematical properties of the linearized large mass scalar field configuration (scalar cloud) with nontrivial coupling to the electromagnetic field of the Kerr black hole. The line element of space-time of spherically symmetric Kerr black hole can be expressed in the form of [13]

$$ds^2 = -g(r)dt^2 + \frac{1}{g(r)}dr^2 + r^2(d\theta^2 + \sin^2\theta d\varphi^2), \tag{1.2}$$

where

$$g(r) = 1 - \frac{2M}{r} + \frac{a^2}{r^2}. \tag{1.3}$$

7

Here M and a are respectively the black-hole mass and its electric charge. The black-hole horizon radii $\{r_+, r_-\}$ are determined by the polynomial equation $h(r = r_\pm) = 0$, which yields

$$r_\pm = M + (M^2 - a^2)^{1/2}. \tag{1.4}$$

The composed Kerr-black-hole-non-minimally-coupled-massive-scalar-field system is characterized by the action

$$S = \int d^4x \sqrt{-g}[R - 2\nabla_\alpha\varphi\nabla^\alpha\varphi - 2\mu^2\varphi^2 - C(\varphi)\mathcal{T}]. \tag{1.5}$$

The coupling function $C(\varphi)$ of the supported massive scalar field configurations is characterized by the universal quadratic behavior [6–9]

$$C(\varphi) = 1 + \alpha\varphi^2 \tag{1.6}$$

in the weak-field regime, where the dimensionless expansion constant α is the physical coupling parameter of the composed black-hole-field theory. We shall henceforth assume $\alpha > 0$.

Substituting into the line element of the curved black-hole spacetime and using the field decomposition [10]

$$\varphi(r, \theta, \varphi) = \sum_{lm} \frac{\psi_{lm}(r)}{r} Y_{lm}(\theta)e^{im\varphi}, \tag{1.7}$$

one finds that the spatial behavior of the static non-minimally coupled massive scalar field configurations, which are supported by the central Kerr black hole, is determined by the ordinary differential equation

$$\frac{d^2\psi}{dy^2} - V\psi = 0, \tag{1.8}$$

where $\mu = y\omega$.

In the next section we shall use analytical techniques in order to determine the discrete resonant spectrum $\{\alpha_n(\mu, l, M, a)\}_{n=0}^{n=\infty}$ of the dimensionless physical parameter α. This resonant spectrum is determined by the Schrödinger-like radial differential equation with the following boundary conditions [9–13]

$$\psi(r = r_+) < \infty; \psi(r \to \infty) \to \frac{1}{r}e^{-\mu r}. \tag{1.9}$$

In the event horizon of the outer black hole and the infinite distance of space, the boundary conditions of these physical excitations correspond to the spatial regular bound state mass scalar field configuration supported by the central rotating black hole.

1.3 The superradiation effect and uncertainty principle

We find the Klein-Gordon equation [13]

$$\Phi_{;\mu}{}^{;\mu} = 0, \tag{1.10}$$

where we defined $\Phi_{;\mu} \equiv (\partial_\mu - ieA_\mu)\Phi$ and e is the charge of the scalar field. We get $A^\mu = \{A_0(x), 0\}$, and $eA_0(x)$ can be equal to μ (where μ is the mass).

$$A_0 \to \begin{cases} 0 & \text{as } x \to -\infty \\ V & \text{as } x \to +\infty \end{cases}. \tag{1.11}$$

With $\Phi = e^{-i\omega t}f(x)$, which is determined by the ordinary differential equation

$$\frac{d^2 f}{dx^2} + (\omega - eA_0)^2 f = 0. \tag{1.12}$$

We see that particles coming from $-\infty$ and scattering off the potential with reflection and transmission amplitudes $\mathcal{R}$ and $\mathcal{T}$ respectively. With these boundary conditions, the solution to behaves asymptotically as

$$f_{\text{in}}(x) = \mathcal{I}e^{i\omega x} + \mathcal{R}e^{-i\omega x}, x \to -\infty, \tag{1.13}$$

$$f_{\text{in}}(x) = \mathcal{T}e^{ikx}, x \to +\infty \tag{1.14}$$

where $k = \pm(\omega - eV)$.

To define the sign of ω and k we must look at the wave's group velocity. We require $\partial\omega/\partial k > 0$, so that they travel from the left to the right in the x–direction and we take $\omega > 0$.

The reflection coefficient and transmission coefficient depend on the specific shape of the potential A_0. However one can easily show that the Wronskian

$$W = \tilde{f}_1 \frac{d\tilde{f}_2}{dx} - \tilde{f}_2 \frac{d\tilde{f}_1}{dx}, \tag{1.15}$$

between two independent solutions, $\tilde{f}_1$ and $\tilde{f}_2$, of is conserved. From the equation on the other hand, if f is a solution then its complex conjugate f^* is another linearly independent solution. We find $|\mathcal{R}|^2 = |\mathcal{I}|^2 - \frac{\omega - eV}{\omega}|\mathcal{T}|^2$. Thus, for $0 < \omega < eV$, it is possible to have superradiant amplification of the reflected current, i.e, $|\mathcal{R}| > |\mathcal{I}|$. There are other potentials that can be completely resolved, which can also show superradiation explicitly.

Classical superradiation effect in the space-time of a steady black hole: we know that $\psi \sim \exp(-i\omega t + im\varphi)$, and the ordinary differential equation

$$\frac{d^2\psi}{dr_*^2} + V\psi = 0. \tag{1.16}$$

We see that particles coming from $-\infty$ and scattering off the potential with reflection and transmission amplitudes $\mathcal{C}$ and $\mathcal{D}$ respectively. With these boundary conditions, the solution to behaves asymptotically as

$$\psi = \begin{cases} Ae^{i\omega_H r_*} + Be^{-i\omega_H r_*}, & r \to r_+ \\ Ce^{i\omega_\infty r_*} + De^{-i\omega_\infty r_*}, & r \to \infty \end{cases}. \tag{1.17}$$

The reflection coefficient and transmission coefficient depend on the specific shape of the potential V. We show that the Wronskian

$$W \equiv \psi \frac{d\bar{\psi}}{dr_*} - \bar{\psi}\frac{d\psi}{dr_*}, \tag{1.18}$$

9

$W(r \to r_+) = 2i\omega_H(|A|^2 - |B|^2)$, $W(r \to \infty) = 2i\omega_\infty(|C|^2 - |D|^2)$ is conserved. We find $|C|^2 - |D|^2 = \frac{\omega_H}{\omega_\infty}(|A|^2 - |B|^2)$. Thus, for $\omega_H/\omega_\infty < 0$, it is possible to have superradiant amplification of the reflected current, i.e, if $|A| = 0$, $|C| > |D|$. There are other potentials that can be completely resolved, which can also show superradiation explicitly.

The principle of joint uncertainty shows that the joint measurement of position and momentum is impossible, that is, the simultaneous measurement of position and momentum can only be an approximate joint measurement, and the error follows the inequality $\Delta x \Delta p \geq 1/2$ (in natural unit system). We find $|\mathcal{R}|^2 = |\mathcal{I}|^2 - \frac{\omega - eV}{\omega}|\mathcal{T}|^2$, and we know that $|\mathcal{R}|^2 \geq -\frac{\omega - eV}{\omega}|\mathcal{T}|^2$ is a necessary condition for the inequality $\Delta x \Delta p \geq 1/2$ to be established. We can pre-set the boundary conditions $eA_0(x) = y\omega$ (which can be $\mu = y\omega$), and we see that when y is relatively large (according to the properties of the boson, y can be very large), $|\mathcal{R}|^2 \geq -\frac{\omega - eV}{\omega}|\mathcal{T}|^2$ may not hold. In the end, we can get $\Delta x \Delta p \geq 1/2$ may not hold. Classical superradiation effect in the space-time of a steady black hole, generalized uncertainty principle may not hold. The same goes for reverse inference.

1.4 New Gravitational coupling equation

We can pre-set the boundary conditions $eA_0(x) = y\omega$ (which can be $\mu = y\omega$) [13, 14], and we see that when y is relatively large (according to the properties of the boson, y can be very large), $|\mathcal{R}|^2 \geq -\frac{\omega - eV}{\omega}|\mathcal{T}|^2$ may not hold. If the boundary conditions of the incident boson are set in advance, the two sides of the probability flow density equation are not equal because of the boundary conditions. Implying a certain probability, this also explains why the no-hair theorem is invalid in quantum effects.

The previous literature [18] indicates that the superradiation effect is a process of entropy subtraction.

Spherical quantum solution in vacuum state.

In this theory, the general relativity theory's field equation is written completely.

$$R_{\mu\nu} - \frac{1}{2}g_{\mu\nu}R = -\frac{8\pi G}{c^4}T_{\mu\nu} \tag{1.19}$$

The Ricci tensor is by $T_{\mu\nu} = 0$ in vacuum state.

$$R_{\mu\nu} = 0 \tag{1.20}$$

The proper time of spherical coordinates is

$$d\tau^2 = A(t,r)dt^2 - \frac{1}{c^2}[B(t,r)dr^2 + r^2 d\theta^2 + r^2 \sin\theta d\varphi^2] \tag{1.21}$$

If we use Eq. (27), we obtain the Ricci-tensor equations.

$$R_{tt} = -\frac{A'}{2B} + \frac{A'B'}{4B^2} - \frac{A'}{Br} + \frac{A'^2}{4AB} + \frac{\ddot{B}}{2B} - \frac{\dot{B}^2}{4B^2} - \frac{\dot{A}\dot{B}}{4AB} = 0 \tag{1.22}$$

$$R_{rr} = \frac{A'}{2A} - \frac{A'^2}{4A^2} - \frac{A'B'}{4AB} - \frac{B'}{Br} - \frac{\ddot{B}}{2A} + \frac{\dot{A}\dot{B}}{4A^2} + \frac{\dot{B}^2}{4AB} = 0, \tag{1.23}$$

$$R_{\theta\theta} = -1 + \frac{1}{B} - \frac{rB'}{2B^2} + \frac{rA'}{2AB} = 0, R_{\varphi\varphi} = R_{\theta\theta}\sin^2\theta = 0,$$

$$R_{tr} = -\frac{\dot{B}}{Br} = 0, R_{t\theta} = R_{t\varphi} = R_{r\theta} = R_{r\varphi} = R_{\theta\varphi} = 0 \tag{1.24}$$

In this time, $' = \frac{\partial}{\partial r}, \cdot = \frac{1}{c}\frac{\partial}{\partial t}$,

$$\dot{B} = 0 \qquad (1.25)$$

We see that,

$$\frac{R_{tt}}{A} + \frac{R_{rr}}{B} = -\frac{1}{Br}\left(\frac{A'}{A} + \frac{B'}{B}\right) = -\frac{(AB)'}{rAB^2} = 0 \qquad (1.26)$$

Hence, we obtain this result.

$$A = \frac{1}{B} \qquad (1.27)$$

If,

$$R_{\theta\theta} = -1 + \frac{1}{B} - \frac{rB'}{2B^2} + \frac{rA'}{2AB} = -1 + \left(\frac{r}{B}\right)' = 0 \qquad (1.28)$$

If we solve the Eq,

$$\frac{r}{B} = r + C \rightarrow \frac{1}{B} = 1 + \frac{C}{r} \qquad (1.29)$$

When r tends to infinity, and we set $C = ye^{-y}$, Therefore,

$$A = \frac{1}{B} = 1 - \frac{y}{r}\Sigma, \Sigma = e^{-y} \qquad (1.30)$$

$$d\tau^2 = \left(1 - \frac{y}{r}\sum\right)dt^2 \qquad (1.31)$$

In this time, if particles' mass are m_i, the fusion energy is e,

$$E = Mc^2 = m_1c^2 + m_2c^2 + \ldots + m_nc^2 + T. \qquad (1.32)$$

1.5 The discrete resonant spectrum of the composed Kerr-black-hole-linearized-massive-scalar-field system: A WKB analysis

In the present section we shall derive a remarkably compact analytical formula for the discrete resonant spectrum $\{\alpha_n(\mu, l, M, a)\}_{n=0}^{n=\infty}$ which characterizes the composed Kerr-black-hole-linearized-massive-scalar-field configurations in the dimensionless large-mass regime [15]

$$M\mu \gg \max\{1, l\}. \qquad (1.33)$$

As we shall now show explicitly, the Schrödinger-like equation, which determines the radial functional behavior of the spatially bounded non-minimally coupled massive scalar field configurations in the kerr black-hole spacetime, is amenable to a WKB analysis in the large-mass regime. In particular, a standard second-order WKB analysis of the Schrödinger-like radial equation yields the well-known discrete quantization condition

$$\int_{y_{t-}}^{y_{t+}} dy\sqrt{-V(y; M, a, l, \mu, \alpha)} = \left(n + \frac{1}{2}\right) \cdot \pi; \quad n = 0, 1, 2, \cdots. \qquad (1.34)$$

11

The two integration boundaries $\{y_{t-}, y_{t+}\}$ of the WKB formula are the classical turning points [with $V(y_{t-}) = V(y_{t+}) = 0$] of the composed charged-black-hole-massive-field binding potential. The resonant parameter n (with $n \in \{0, 1, 2, \cdots\}$) characterizes the infinitely large discrete resonant spectrum $\{\alpha_n(\mu, l, M, a)\}_{n=0}^{n=\infty}$ of the black-hole-field system.

Using the relation between the radial coordinates y and r, one can express the WKB resonance equation in the form

$$\int_{r_{t-}}^{r_{t+}} dr \frac{\sqrt{-V(r; M, a, l, \mu, \alpha)}}{h(r)} = \left(n + \frac{1}{2}\right) \cdot \pi; \quad n = 0, 1, 2, \cdots, \tag{1.35}$$

where the two polynomial relations

$$1 - \frac{2M}{r_{t-}} + \frac{a^2}{r_{t-}^2} = 0 \tag{1.36}$$

and

$$\frac{l(l+1)}{r_{t+}^2} + \frac{2M}{r_{t+}^3} - \frac{2a^2}{r_{t+}^4} - \frac{\alpha a^2}{r_{t+}^4} = 0 \tag{1.37}$$

determine the radial turning points $\{r_{t-}, r_{t+}\}$ of the composed black-hole-field binding potential.

We shall now prove that the WKB resonance equation can be studied *analytically* in the regime of large field masses. To this end, it proves useful to define the dimensionless physical quantities

$$x \equiv \frac{r - r_+}{r_+}; \tau \equiv \frac{r_+ - r_-}{r_+}, \tag{1.38}$$

in terms of which the composed black-hole-massive-field interaction term has the form of a binding potential well,

$$V[x(r)] = -\tau \left(\frac{\alpha a^2}{r_+^4} - \mu^2\right) \cdot x + \left[\frac{\alpha a^2 (5r_+ - 6r_-)}{r_+^5} - \mu^2 \left(1 - \frac{2r_-}{r_+}\right)\right] \cdot x^2 + O(x^3), \tag{1.39}$$

in the near-horizon region

$$x \ll \tau. \tag{1.40}$$

From the near-horizon expression of the black-hole-field binding potential, one obtains the dimensionless expressions

$$x_{t-} = 0 \tag{1.41}$$

and

$$x_{t+} = \tau \cdot \frac{\frac{\alpha a^2}{r_+^4} - \mu^2}{\frac{\alpha a^2 (5r_+ - 6r_-)}{r_+^5} - \mu^2 \left(1 - \frac{2r_-}{r_+}\right)} \tag{1.42}$$

for the classical turning points of the WKB integral relation.

We find that our analysis is valid in the regime below

$$\alpha \simeq \frac{\mu^2 r_+^4}{a^2}, \tag{1.43}$$

in which case the near-horizon binding potential and its outer turning point can be approximated by the remarkably compact expressions

$$V(x) = -\tau \left[\left(\frac{\alpha a^2}{r_+^4} - \mu^2 \right) \cdot x - 4\mu^2 \cdot x^2 \right] + O(x^3) \tag{1.44}$$

and

$$x_{t+} = \frac{1}{4} \left(\frac{\alpha a^2}{\mu^2 r_+^4} - 1 \right). \tag{1.45}$$

In addition, one finds the near-horizon relation

$$g(x) = \tau \cdot x + (1 - 2\tau) \cdot x^2 + O(x^3). \tag{1.46}$$

We know that

$$\frac{1}{\sqrt{\tau}} \int_0^{x_{t+}} dx \sqrt{\frac{\frac{\alpha a^2}{r_+^2} - \mu^2 r_+^2}{x} - 4\mu^2 r_+^2} = \left(n + \frac{1}{2} \right) \cdot \pi; \quad n = 0, 1, 2, \cdots. \tag{1.47}$$

Defining the dimensionless radial coordinate

$$z \equiv \frac{x}{x_{t+}}, \tag{1.48}$$

one can express the WKB resonance equation in the mathematically compact form

$$\frac{2\mu r_+ x_{t+}}{\sqrt{\tau}} \int_0^1 dz \sqrt{\frac{1}{z} - 1} = \left(n + \frac{1}{2} \right) \cdot \pi; \quad n = 0, 1, 2, \cdots, \tag{1.49}$$

which yields the relation

$$\frac{\mu r_+ x_{t+}}{\sqrt{\tau}} = n + \frac{1}{2}; n = 0, 1, 2, \cdots \tag{1.50}$$

We find the discrete resonant spectrum

$$\alpha_n = \frac{\mu^2 r_+^4}{a^2} \left[1 + \frac{4\sqrt{\tau}}{\mu r_+} \left(n + \frac{1}{2} \right) \right]; \quad n = 0, 1, 2, \cdots \tag{1.51}$$

for the dimensionless coupling parameter of the composed kerr-black-hole-non-minimally-coupled-linearized-massive-scalar-field configurations in the regime $\mu r_+ \gg 1$ of large field masses. The analytically derived relation can also be written as the discrete resonant formula

$$(\mu r_+)_n = \sqrt{\alpha \frac{r_-}{r_+}} - \sqrt{\frac{r_+ - r_-}{r_+}} \cdot (2n + 1) \text{ for } \alpha \gg 1 \tag{1.52}$$

for the dimensionless mass parameter which characterizes the non-minimally coupled massive scalar field clouds in the large-coupling $\alpha \gg 1$ regime.

13

1.6 Summary and Discussion: The y limit

Interestingly, it has been demonstrated numerically in [16, 17] that the dimensionless physical parameter α *diverges* in the y limit, where the physical parameter y is defined by the dimensionless relation

$$\frac{\alpha}{y} \equiv \mu^2 r_+^2.$$
(1.53)

Here the critical parameter y is given by the simple relation

$$y \equiv \frac{r_+^2}{a^2}.$$
(1.54)

The limit y of the incident particle under the Hawking radiation of the preset boundary $(y\mu = \omega)$:

$$y \equiv \frac{\alpha}{\mu^2 r_+^2}.$$
(1.55)

To calculate the life of a black hole, we must assume that the black hole only generates external radiation and does not absorb matter. Then all the mass lost by the black hole is emitted in the form of radiated photons. Then the radiation life is equal to

$$\frac{dMc^2}{dt} = -Aj_{u^\circ}.$$
(1.56)

If you know the surface area of the black hole and the radiant energy flux density, you can calculate the life of the black hole. For

$$r \to \infty (\mathrm{r}_* \to \infty) \Rightarrow R_{\mathrm{lm}} \sim \frac{1}{r} e^{-\sqrt{\mu^2 - \omega^2} r_*},$$
(1.57)

$$j_{\mathrm{out}} = \frac{-i\hbar}{2m} \left(\psi_{\mathrm{out}} \frac{\partial}{\partial r} \psi_{\mathrm{out}}^* - \psi_{\mathrm{out}}^* \frac{\partial}{\partial r} \psi_{\mathrm{out}} \right) = v |\psi_{\mathrm{out}}^2| = \frac{1}{b^2} \exp\left[-\frac{2}{\hbar} (\mathrm{Im}\, S_0) \right].$$
(1.58)

When j_{out} takes the maximum value, we can know the maximum lifetime of the Kerr black hole in that situation.

Bibliography

[1] J. D. Bekenstein, Phys. Rev. D 5, 1239 (1972).

[2] C. A. R. Herdeiro and E. Radu, Int. J. Mod. Phys. D 24, 1542014 (2015).

[3] S. Hod, Phys. Lett. B 713, 505 (2012); S. Hod, Phys. Lett. B 718, 1489 (2013) [arXiv: 1304.6474]; S. Hod, Phys. Rev. D 91, 044047 (2015) [arXiv: 1504.00009].

[4] A. E. Mayo and J. D. Bekenstein and, Phys. Rev. D 54, 5059 (1996).

[5] S. Hod, Phys. Lett. B 771, 521 (2017); S. Hod, Phys. Rev. D 96, 124037 (2017).

[6] C. A. R. Herdeiro, E. Radu, N. Sanchis-Gual, and J. A. Font, Phys. Rev. Lett. 121, 101102 (2018).

[7] P. G. S. Fernandes, C. A. R. Herdeiro, A. M. Pombo, E. Radu, and N. Sanchis-Gual, Class. Quant. Grav. 36, 134002 (2019) [arXiv: 1902.05079].

[8] D. C. Zou and Y. S. Myung, Phys. Rev. D 100, 124055 (2019).

[9] P. G. S. Fernandes, arXiv: 2003.01045.

[10] Here Q is the electric charge of the charged black-hole spacetime. We shall henceforth assume that $Q > 0$ without loss of generality.

[11] S. Hod, Phys. Rev. D 86, 104026 (2012) [arXiv: 1211.3202]; S. Hod, The Euro. Phys. Journal C 73, 2378 (2013) [arXiv: 1311.5298]; S. Hod, Phys. Rev. D 90, 024051 (2014) [arXiv: 1406.1179].

[12] C. A. R. Herdeiro and E. Radu, Phys. Rev. Lett. 112, 221101 (2014).

[13] Brito, Richard, Vitor Cardoso, and Paolo Pani. Superradiance. Springer International Publishing, 2020.

[14] Chen, Wen-Xiang, and Zi-Yang Huang. "Superradiant stability of the kerr black hole." International Journal of Modern Physics D 29.01 (2020): 2050009.

[15] Hod, Shahar. "Reissner-Nordström black holes supporting nonminimally coupled massive scalar field configurations." Physical Review D 101.10 (2020): 104025.

[16] Chen, Wen-Xiang. "The possibility of the no-hair theorem being violated." Available at SSRN 3569639 (2020).

[17] Chen, Wen-Xiang. "Uncertainty principle and super-radiance." Available at SSRN 3683008 (2020).

[18] Yi, Sangwha. "Spherical Solution of Classical Quantum Gravity." Available at SSRN 3508075 (2019).

[19] Luca Giacomelli. "Superradiant phenomena Lessons from and for Bose-Einstein condensates." https://iris.unitn.it. phd unitn luca giacomelli.

Chapter 2

Evidence for loop quantum gravity

Abstract: This article points out that when the boundary condition $\frac{T}{T_c} = z$ (when z is a complex number) is preset, bosons can produce Bose condensation without an energy layer. Under Bose condensation, incident waves may condense in various black holes in the theory of loop quantum gravity. At that time, potential barriers will be generated near the horizon of various black holes, and we believe that these black holes will also produce super-radiation phenomena (this article uses the natural unit system). We assume that the simple loop quantum gravity theoretical model that can produce superradiation phenomena that does not exist in the traditional theory provides experimental evidence for loop quantum gravity.

Keywords: quaternion, Schwarzschild black hole, temperature

2.1 Introduction

Loop quantum gravity theory, the English alias loop gravity, quantum geometry; the quantum gravity theory developed by Abbe Ashitika, Lee Smalling, Carlo Rowley, etc., is the same as string theory It is the most successful theory of quantizing gravity today.

The theory of using the perturbation theory of quantum field theory to realize the quantization of the theory of gravity cannot be renormalized. If you start with general relativity while advocating that space-time has only four dimensions, you can turn general relativity into a theory similar to gauge field theory. The basic regular variable is the Ashitika-Barbero connection instead of the metric tensor, and then the connection is defined The translation operator and the flux variable are the basic variables to achieve quantization.

Under this theory, the description of space-time is background-independent, and a spin network woven by relational loops is paved with space-time geometry. The length of each side in the network is the Planck length. The loop does not exist in space and time, but the geometry of space and time is defined in the way of loops and twists. On the Planck scale, the geometry of space-time is full of random quantum fluctuations, so spin networks are also called spin bubbles. Under this theory, time and space are discrete.

Most string theorists believe that it is impossible to quantize gravity in 3+1 dimensional spacetime without producing artificial products related to matter and energy. However, the matter-related artifacts predicted by string theory have not been proven whether they are really different from the actual observed matter. However, if loop quantum gravity succeeds in becoming a quantum theory of gravity, the known matter field must be added to this theory "after the fact" instead of appearing spontaneously from the theory. Lee Smalling, one of the founders of loop quantum gravity, has considered the possibility that string theory and loop quantum gravity may be two different approximations of the ultimate theory.

Simple spin network morphology used in loop quantum theory The current claimed successes of loop quantum gravity include:

It is a non-perturbative quantization of 3-dimensional space geometry, with quantized area and volume operators. It includes the calculation of black hole entropy. It is another feasible theory besides string theory, but it only involves the quantization of gravity (that is, non-universal theory). However, such claims have not yet been fully accepted. Although many of the core results of loop quantum gravity are derived from rigorous mathematical physics, their physical interpretations are still mostly speculative. Loop quantum gravity is likely to be an improvement to gravity or geometry; for example, the entropy calculation in (2) is actually done for a form of "hole", which may or may not be a black hole.

Other schemes of quantum gravity, such as the spin foam model, are closely related to looping quantum gravity.

The two most important assumptions of loop quantum gravity are

General Covariance-The laws of physics can be expressed in any coordinate system, which is also the basic assumption of general relativity. Background independence-There is no independent and unchanging metric, coordinate system, etc. that can be used as a background. Loop quantum gravity also assumes that the basic principles of quantum theory are correct. For example, generalized covariance theories include general relativity, non-generalized covariation theories include special relativity (special covariance), non-background independent theories include Newtonian mechanics (assuming an independent and unchanging time axis), and special relativity (its background It is the Minkowski space, the background metric is the Minkowski metric), the equation of electrons moving in the background electromagnetic field, etc., background-independent theories include general relativity, and the value of the metric tensor is completely determined by the theory.

In the reference frame of a specific superradiation (Loop quantum gravity theory), in its self-reference frame, γ modifies the structure. In the free space where $\gamma < 1$, the electron is not a stable particle, and its wave function diverges and grows. However, in the binding state, this divergence is controlled by the binding potential, and γ can take a value less than 1.

If the boundary conditions are preset, the boundary conditions $\frac{T}{T_c} = z$ (when z is plural), the effective action form satisfies the effective action form of Hawking radiation, and is not necessarily at the boundary of event horizon. The general formulation of the black hole temperature is [4] (At that time there was a special solution), finishing the integral we obtain

$$\operatorname{Im} T = -\frac{2}{\pi} \left(\frac{1}{r_f'^2} - \frac{1}{r_i'^2} \right) = -\frac{1}{2} \Delta T_{\mathrm{CH}} \tag{2.1}$$

This article points out that when the boundary conditions $\frac{T}{T_c} = z$ (when z is plural) are preset,

18

bosons can produce Bose condensation without an energy layer. Under Bose condensation, the incident wave may be condensed in various black holes in the theory of loop quantum gravity. At that time, potential barriers can be generated near the event horizon of various black holes, and we believe that those black holes will also produce superradiation phenomena (this article adopts the natural unit system). We assume that the simple theoretical model of loop quantum gravity, which can produce super-radiation phenomena that do not exist in traditional theory, provides experimental evidence of loop quantum gravity.

2.2   The superradiation effect and uncertainty principle

We find the Klein-Gordon equation [10]

$$\Phi_{;\mu}^{;\mu} = 0 \,, \tag{2.2}$$

where we defined $\Phi_{;\mu} \equiv (\partial_\mu - ieA_\mu)\Phi$ and e is the charge of the scalar field. We get $A^\mu = \{A_0(x), 0\}$, and $eA_0(x)$ can be equal to μ (where μ is the mass).

$$A_0 \to \begin{cases} 0 & \text{as } x \to -\infty \\ V & \text{as } x \to +\infty \end{cases}. \tag{2.3}$$

With $\Phi = e^{-i\omega t} f(x)$, which is determined by the ordinary differential equation

$$\frac{\mathrm{d}^2 f}{\mathrm{d}x^2} + (\omega - eA_0)^2 f = 0. \tag{2.4}$$

We see that particles coming from $-\infty$ and scattering off the potential with reflection and transmission amplitudes $\mathcal{R}$ and $\mathcal{T}$ respectively. With these boundary conditions, the solution to behaves asymptotically as

$$f_{\text{in}}(x) = \mathcal{I}e^{i\omega x} + \mathcal{R}e^{-i\omega x}, \quad x \to -\infty, \tag{2.5}$$

$$f_{\text{in}}(x) = \mathcal{T}e^{ikx}, \quad x \to +\infty \tag{2.6}$$

where $k = \pm(\omega - eV)$.

To define the sign of ω and k we must look at the wave's group velocity. We require $\partial\omega/\partial k > 0$, so that they travel from the left to the right in the x-direction and we take $\omega > 0$.

The reflection coefficient and transmission coefficient depend on the specific shape of the potential A_0. However one can easily show that the Wronskian

$$W = \tilde{f}_1 \frac{\mathrm{d}\tilde{f}_2}{\mathrm{d}x} - \tilde{f}_2 \frac{\mathrm{d}\tilde{f}_1}{\mathrm{d}x}, \tag{2.7}$$

between two independent solutions, $\tilde{f}_1$ and $\tilde{f}_2$, of is conserved. From the equation on the other hand, if f is a solution then its complex conjugate f^* is another linearly independent solution. We find $|\mathcal{R}|^2 = |\mathcal{I}|^2 - \frac{\omega - eV}{\omega}|\mathcal{T}|^2$. Thus, for $0 < \omega < eV$, it is possible to have superradiant amplification of the reflected current, i.e, $|\mathcal{R}| > |\mathcal{I}|$. There are other potentials that can be completely resolved, which can also show superradiation explicitly.

Classical superradiation effect in the space-time of a steady black hole: we know that $\psi \sim \exp(-i\omega t + im\varphi)$, and the ordinary differential equation

$$\frac{d^2\psi}{dr_*^2} + V\psi = 0. \tag{2.8}$$

We see that particles coming from $-\infty$ and scattering off the potential with reflection and transmission amplitudes $\mathcal{C}$ and $\mathcal{D}$ respectively. With these boundary conditions, the solution to behaves asymptotically as

$$\psi = \begin{cases} Ae^{i\omega_H r_*} + Be^{-i\omega_H r_*}, & r \to r_+ \\ Ce^{i\omega_\infty r_*} + De^{-i\omega_\infty r_*}, & r \to \infty \end{cases}. \tag{2.9}$$

The reflection coefficient and transmission coefficient depend on the specific shape of the potential V. We show that the Wronskian

$$W \equiv \psi\frac{d\bar{\psi}}{dr_*} - \bar{\psi}\frac{d\psi}{dr_*}, \tag{2.10}$$

$W(r \to r_+) = 2i\omega_H(|A|^2 - |B|^2), W(r \to \infty) = 2i\omega_\infty(|C|^2 - |D|^2)$ is conserved. We find $|C|^2 - |D|^2 = \frac{\omega_H}{\omega_\infty}(|A|^2 - |B|^2)$. Thus, for $\omega_H/\omega_\infty < 0$, it is possible to have superradiant amplification of the reflected current, i.e, if $|A| = 0$, $|C| > |D|$. There are other potentials that can be completely resolved, which can also show superradiation explicitly.

The principle of joint uncertainty shows that the joint measurement of position and momentum is impossible, that is, the simultaneous measurement of position and momentum can only be an approximate joint measurement, and the error follows the inequality $\Delta x\Delta p \geq 1/2$ (in natural unit system). We find $|\mathcal{R}|^2 = |\mathcal{I}|^2 - \frac{\omega-eV}{\omega}|\mathcal{T}|^2$, and we know that $|\mathcal{R}|^2 \geq -\frac{\omega-eV}{\omega}|\mathcal{T}|^2$ is a necessary condition for the inequality $\Delta x\Delta p \geq 1/2$ to be established. We can pre-set the boundary conditions $eA_0(x) = y\omega$ (which can be $\mu = y\omega$), and we see that when y is relatively large (according to the properties of the boson, y can be very large), $|\mathcal{R}|^2 \geq -\frac{\omega-eV}{\omega}|\mathcal{T}|^2$ may not hold. In the end, we can get $\Delta x\Delta p \geq 1/2$ may not hold. Classical superradiation effect in the space-time of a steady black hole, generalized uncertainty principle may not hold. The same goes for reverse inference.

We can pre-set the boundary conditions $eA_0(x) = -y\omega$ (which can be $\mu = y\omega$). [11]

$$V_1(r) = -\frac{Ze^2}{4\pi\epsilon_o r}$$

$$\Delta V(r) = V_1(r) - \int_\infty^r u^{f*}(x)u^f(x)dx \tag{2.11}$$

$$\nabla^2\Phi - \frac{1}{c^2}\frac{\partial^2\Phi}{\partial t^2} = \frac{2m_o}{\hbar^2}\left\{-i\hbar\frac{\partial\Phi}{\partial t} + V\left(1 + \frac{V}{2m_oc^2}\right)\Phi\right\}$$

$$i\hbar\frac{\partial\Phi}{\partial t} = -\frac{\hbar^2}{m_o(1+\gamma)}\nabla^2\Phi + \frac{2V}{1+\gamma}\left(1 + \frac{V}{2m_oc^2}\right)\Phi \tag{2.12}$$

In the reference frame of a specific superradiation (Loop quantum gravity theory), in its self-reference frame, γ modifies the structure. In the free space where $\gamma < 1$, the electron is not a stable particle, and its wave function diverges and grows. However, in the binding state, this divergence is controlled by the binding potential, and γ can take a value less than 1.

2.3 Similar to the ADS space

We can pre-set the boundary conditions $eA_0(x) = -y\omega$ (which can be $\mu = y\omega$). If the boundary conditions of the incident boson are set in advance, the two sides of the probability flow density equation are not equal because of the boundary conditions. Implying a certain probability, this also explains why the no-hair theorem is invalid in quantum effects.

The previous literature [10] indicates that the superradiation effect is a process of entropy subtraction.

Spherical quantum solution in vacuum state.

In this theory, the general relativity theory's field equation is written completely.

$$R_{\mu\nu} - \frac{1}{2} g_{\mu\nu} R = -\frac{8\pi G}{c^4} T_{\mu\nu} \tag{2.13}$$

The Ricci tensor is by $T_{\mu\nu} = 0$ in vacuum state.

$$R_{\mu\nu} = 0 \tag{2.14}$$

The proper time of spherical coordinates is

$$d\tau^2 = A(t,r)dt^2 - \frac{1}{c^2}[B(t,r)dr^2 + r^2 d\theta^2 + r^2 \sin\theta d\varphi^2] \tag{2.15}$$

We obtain the Ricci-tensor equations.

$$R_{tt} = -\frac{A'}{2B} + \frac{A'B'}{4B^2} - \frac{A'}{Br} + \frac{A'^2}{4AB} + \frac{\ddot{B}}{2B} - \frac{\dot{B}^2}{4B^2} - \frac{\dot{A}\dot{B}}{4AB} = 0 \tag{2.16}$$

$$R_{rr} = \frac{A'}{2A} - \frac{A'^2}{4A^2} - \frac{A'B'}{4AB} - \frac{B'}{Br} - \frac{\ddot{B}}{2A} + \frac{\dot{A}\dot{B}}{4A^2} + \frac{\dot{B}^2}{4AB} = 0, \tag{2.17}$$

$$R_{\theta\theta} = -1 + \frac{1}{B} - \frac{rB'}{2B^2} + \frac{rA'}{2AB} = 0, \ R_{\varphi\varphi} = R_{\theta\theta}\sin^2\theta = 0, \ R_{tr} = -\frac{\dot{B}}{Br} = 0,$$

$$R_{t\theta} = R_{t\varphi} = R_{r\theta} = R_{r\varphi} = R_{\theta\varphi} = 0 \tag{2.18}$$

In this time, $' = \frac{\partial}{\partial r}, \cdot = \frac{1}{c}\frac{\partial}{\partial t}$,

$$\dot{B} = 0 \tag{2.19}$$

We see that,

$$\frac{R_{tt}}{A} + \frac{R_{rr}}{B} = -\frac{1}{Br}\left(\frac{A'}{A} + \frac{B'}{B}\right) = -\frac{(AB)'}{rAB^2} = 0 \tag{2.20}$$

Hence, we obtain this result.

$$A = \frac{1}{B} \tag{2.21}$$

If,

$$R_{\theta\theta} = -1 + \frac{1}{B} - \frac{rB'}{2B^2} + \frac{rA'}{2AB} = -1 + \left(\frac{r}{B}\right)' = 0 \tag{2.22}$$

21

If we solve the Eq,

$$\frac{r}{B} = r + C \rightarrow \frac{1}{B} = 1 + \frac{C}{r} \tag{2.23}$$

When r tends to infinity, and we set $C = -ye^y$, Therefore,

$$A = \frac{1}{B} = 1 + \frac{y}{r}\Sigma, \Sigma = e^y \tag{2.24}$$

$$d\tau^2 = \left(1 + \frac{y}{r}\sum\right)dt^2 \tag{2.25}$$

In this time, if particles' mass are m_i, the fusion energy is e,

$$E = Mc^2 = m_1c^2 + m_2c^2 + \ldots + m_nc^2 + e \tag{2.26}$$

The effect after the preset boundary is similar to that of Ads cosmological constant.

2.4 Thermodynamic phase transition

Thermodynamic phase transition. - The state equation of a charged AdS black hole displays a vdW-like thermodynamic behavior. The SBH-LBH coexistence curve has a parametric form [1]

$$\frac{P}{P_c} = \sum_i a_i \left(\frac{T}{T_c}\right)^i. \tag{2.27}$$

The concept of entropy was proposed by the German physicist Clausius in 1865. Kjeldahl defines the increase and decrease of entropy in a thermodynamic system: the total amount of heat used at a constant temperature in a reversible process, and can be expressed as:

$$\Delta S = \frac{\Delta Q}{T} \tag{2.28}$$

if $\frac{T}{T_c} = z$, when z is plural. The Laurent series of the function $f(z)$ with respect to point c is given by:

$$f(z) = \sum_{n=-\infty}^{\infty} a_n(z - c)^n \tag{2.29}$$

It is defined by the following curve integral, which is a generalization of the Cauchy integral formula:

$$a_n = \frac{1}{2\pi i} \oint_\gamma \frac{f(z)dz}{(z - c)^{n+1}} \tag{2.30}$$

Since the algebra of real quaternions is the only fourdimensional division algebra, we introduce the fourdimensional quaternion manifold, [2]

$$\tau^4 = (\hat{\tau}_0, \vec{\tau}_1, \vec{\tau}_2, \vec{\tau}_3) = (\hat{i}_0\tau_0, \vec{i}_1\tau_1, \vec{i}_2\tau_2, \vec{i}_3\tau_3) \tag{2.31}$$

$$\begin{cases} \hat{i}_0\hat{i}_0 = \hat{i}_0 = 1 \\ \vec{i}_1\vec{i}_1 = \vec{i}_2\vec{i}_2 = \vec{i}_3\vec{i}_3 = \vec{i}_1\vec{i}_2\vec{i}_3 = -\hat{i}_0 = -1 \\ \vec{i}_1\vec{i}_2 = \vec{i}_3, \vec{i}_2\vec{i}_3 = \vec{i}_1, \vec{i}_3\vec{i}_1 = \vec{i}_2, \\ \vec{i}_2\vec{i}_1 = -\vec{i}_3, \vec{i}_3\vec{i}_2 = -\vec{i}_1, \vec{i}_1\vec{i}_3 = -\vec{i}_2 \end{cases} \tag{2.32}$$

22

$$t = \left(\hat{\iota}_0 t_0, \vec{\iota}_1 \frac{x_1}{c}, \vec{\iota}_2 \frac{x_2}{c}, \vec{\iota}_3 \frac{x_3}{c} \right) \tag{2.33}$$

$$\begin{cases} t = t \left(\frac{t_0}{t}, \frac{\vec{v}}{c} \right) = t(\cos\theta, \vec{\iota}\sin\theta) = t \exp(\vec{\iota}\theta) \\ \bar{\iota} = t \left(\frac{t_0}{t}, -\frac{\vec{v}}{c} \right) = t(\cos\theta, -\vec{\iota}\sin\theta) = t \exp(-\vec{\iota}\theta) \end{cases} \tag{2.34}$$

$$\begin{cases} t = \frac{t_0}{\sqrt{1-\frac{v^2}{c^2}}} \exp(\vec{\iota}\theta) \\ \bar{\iota} = \frac{t_0}{\sqrt{1-\frac{v^2}{c^2}}} \exp(-\vec{\iota}\theta) \end{cases} \tag{2.35}$$

2.5 Schwarzschild- AdS black hole produce super-radiation phenomena

To describe tunneling, we need a coordinate system that is regular at the horizons; particularly convenient are Painlevé coordinates. The line element of the Schwarzschild Ads space-time is

$$ds^2 = -\left(1 - \frac{2m}{r} - \frac{\lambda}{3}r^2 \right) dt_s^2 + \left(1 - \frac{2m}{r} - \frac{\lambda}{3}r^2 \right)^{-1} dr^2 + r^2 d\Omega^2$$

where m is the mass of the black hole and λ is the cosmological constant.

When the tunneling takes place at the ADS cosmological horizon, the particle is propagating from larger to smaller r, and the mass of the black hole increases from m to $m + \omega'$. If the self-gravitation is included, the form of $\dot{r}$ become [12]

$$\dot{r} = \frac{1}{2r} \sqrt{\frac{\lambda r}{3}} \frac{\left(r^3 - \frac{3}{\lambda}r + \frac{6(m+\omega')}{\lambda} \right)}{\sqrt{r^3 + \frac{6(m+\omega')}{\lambda}}} = \frac{1}{2r} \sqrt{\frac{\lambda r}{3}} \frac{(r - r'_-)(r - r'_h)(r - r'_c)}{\sqrt{r^3 + \frac{6(m+\omega')}{\lambda}}} \tag{2.36}$$

where

$$r'_c = 2\sqrt{\frac{1}{\lambda}} \cos\frac{\phi'}{3}$$

$$r'_h = -2\sqrt{\frac{1}{\lambda}} \cos\left(\frac{\phi'}{3} + \frac{\pi}{3} \right)$$

$$r'_- = -2\sqrt{\frac{1}{\lambda}} \cos\left(\frac{\phi'}{3} - \frac{\pi}{3} \right) \tag{2.37}$$

with ϕ' satisfies

$$\cos\phi' = -3(m + \omega')\sqrt{\lambda} \tag{2.38}$$

For a positive-energy s-wave, the imaginary part of the action is

$$\text{Im} \, T = \int_{}^{r'_f} p_r dr = \int_{}^{r'_f} \int_{}^{p_r} dp'_r dr \tag{2.39}$$

$$\text{Im} \, T = \text{Im} \int_{r'_i}^{r'_f} \int_0^{\omega} \dot{r} d\omega' dr \tag{2.40}$$

we have

$$\text{Im} \, T = \text{Im} \int_{r'_i}^{r'_f} \int_0^{\omega} \frac{1}{2r} \sqrt{\frac{\lambda r}{3}} \frac{(r - r'_-)(r - r'_h)(r - r'_c)}{\sqrt{r^3 + \frac{6(m+\omega')}{\lambda}}} d\omega' dr \tag{2.41}$$

23

Finally, finishing the integral we obtain

$$\operatorname{Im} T = -\frac{2}{\pi}\left(\frac{1}{r_f'^2} - \frac{1}{r_i'^2}\right) = -\frac{1}{2}\Delta T_{\text{CH}} \tag{2.42}$$

$$|C|^2 \sim \exp[-2\operatorname{Im} T] = e^{\Delta T_{\text{CH}}} \tag{2.43}$$

It is also consistent with an underlying unitary theory. We see that under the background of the Schwarzschild-ADS black hole, $|C| > |\mathcal{D}|$. At that time, there is a potential barrier near the horizon. We know that the Schwarzschild-ADS black hole can be superradiation at that time (At that time there was a special solution).

2.6 Summary

This article points out that when the boundary conditions $\frac{T}{T_c} = z$ (when z is plural) are preset, bosons can produce Bose condensation without an energy layer. Under Bose condensation, the incident wave may be condensed in various black holes in the theory of loop quantum gravity. At that time, potential barriers can be generated near the event horizon of various black holes, and we believe that those black holes will also produce superradiation phenomena (this article adopts the natural unit system). We assume that the simple theoretical model of loop quantum gravity, which can produce super-radiation phenomena that do not exist in traditional theory, provides experimental evidence of loop quantum gravity.

Acknowledgements

I would like to thank Jing-Yi Zhang. This work is partially supported by National Natural Science Foundation of China (No. 11873025).

Bibliography

[1] Wei, Shao-Wen, and Yu-Xiao Liu. "Clapeyron equations and fitting formula of the coexistence curve in the extended phase space of charged AdS black holes." Physical Review D 91.4 (2015): 044018.

[2] Ariel, Viktor. "Quaternion Space-Time and Matter." arXiv preprint arXiv: 2106.06394 (2021).

[3] Murata K, Soda J. Hawking Radiation from Rotating Black Holes and Gravitational Anomalies[J]. Physical Review D, 2006, 74(4): 200-206.

[4] Medved, A. J. M., and Elias C. Vagenas. "On Hawking radiation as tunneling with logarithmic corrections." Modern Physics Letters A 20.23 (2005): 1723-1728.

[5] M. K. Parikh, F. Wilczek, *Phys. Rev. Lett.*, 85, 5042(2000) [arxiv: hep-th/9907001].

[6] M. K. Parikh, *Int. J. Mod. Phys.* D 13, 2355(2004) [arXiv: hep-th/0405160].

[7] M. K. Parikh, arXiv: hep-th/0402166.

[8] S. Hemming, E. Keski-Vakkuri, *Phys. Rev. D64*, 044006(2001).

[9] Zhang, Jingyi. "Black hole quantum tunnelling and black hole entropy correction." Physics Letters B 668.5 (2008): 353-356.

[10] Brito, Richard, Vitor Cardoso, and Paolo Pani. Superradiance. Springer International Publishing, 2020.

[11] John P. Wallace, Michael J. Wallace. https://rxiv.org/abs/2103.0026.

[12] Zhang, Jingyi, and Zheng Zhao. "Massive particles' black hole tunneling and de Sitter tunneling." Nuclear Physics B 725.1-2 (2005): 173-180.

Chapter 3

Schwarzschild black hole can also produce super-radiation phenomena and the cosmic censorship conjecture may be violated

Abstract: This article points out that when the boundary conditions $\frac{T}{T_c} = z$ (when z is plural) are preset, bosons can produce Bose condensation without an energy layer. Under Bose condensation, the incident wave may condense in the Schwarzschild black hole. At that time, the Schwarzschild black hole event horizon Potential barriers can be generated nearby, and we think that Schwarzschild black holes can also generate superradiation phenomena (This article uses the natural unit system). This implies that the cosmic censorship conjecture may be violated.

Keywords: quaternion, Schwarzschild black hole, temperature

3.1   Introduction

In 2000, Parikh and Wilczek proposed a method to calculate the emissivity of particles passing through the event horizon. They treat Hawking radiation as a tunneling process and use the WKB method. In this way, a correction spectrum accurate to a first-order approximation is given. Their results are considered to be consistent with the basic Mozheng theory. According to this method, a large number of stationary or stationary rotating black holes have been studied, and the same result has been obtained, that is, Hawking radiation is no longer pure thermal radiation, which satisfies the Universe Theory and the conservation of information. But in all these documents, the entropy of black holes only contains Bekenstein-Hawking entropy. If the quantum correction of entropy is considered, does the emission process still conform to the monolithic theory? At present, with regard to quantum correction, different models and methods correspond to different results.

Gas, liquid, and solid are the three basic forms of substances that are well known to people. For example, water molecules can appear as ice, water, and water vapor as the temperature increases. Further scientists discovered that as the temperature of the material continues to rise, a plasma state will appear. An interesting question is how to keep the substance cold, such as close to absolute zero ($-273.16°C$). Under such extremely low temperature, what kind of strange state will the matter appear? In fact, as early as 1924, the famous physicists Bose and Einstein gave the answer to the question. Particles with integer spins, including photons, gluons, and nuclei composed of an even number of nucleons, and alkali metal atoms, are all called bosons. At that time, the Indian physicist Bose proposed a new method of photon statistics. Einstein extended it to an ideal gas with mass, and theoretically predicted that there is no interacting boson, and it will be at its lowest value at a certain temperature. The energy quantum state suddenly condenses, and when it reaches a considerable amount, it is in the Bose-Einstein condensate (BEC).

The Kerr-black hole-scalar field mirror system first designed by Press and Teukolsky has attracted the attention of physicists in the past four decades. From the literature on the numerical study of this kind of rotating black hole bomb, we know an interesting fact, that is, if the reflector is too close to the black hole horizon, super radiation will produce a stabilizing effect. And we know that black holes need to meet the conditions to produce classical superradiation instability:

(1) The incident perturbation field is the Bose field;

(2) The black hole is rotating or charged;

(3) Meet the super-radiation conditions $0 < \omega < m\omega_H$.

(4) The existence of reflective mirrors.

Among them (1), (2), (3) are the conditions for generating super radiation. In 1972, Press and Teukolsky proposed that It is possible to add a mirror to the outside of a black hole to make a black hole bomb (according to the current explanation, this is a scattering process involving classical mechanics and quantum mechanics. Regge and Wheeler proved that the spherically symmetric Schwarzschild black hole is stable under disturbance. Due to the significant influence of super radiation, the stability of rotating black holes is more complicated. Superradiation effects can occur in classical and quantum scattering processes. When a boson wave hits a rotating black hole, if certain conditions are met, the black hole may be as stable as a Schwarzschild black hole. When a boson wave hits a rotating black hole, if the frequency range of the wave is under superradiation conditions, the wave reflected by the event horizon will be amplified.

If the boundary conditions are preset, the boundary conditions $\frac{T}{T_c} = z$ (when z is plural), the effective action form satisfies the effective action form of Hawking radiation, and is not necessarily at the boundary of event horizon. The general formulation of the black hole temperature is [4](At that time there was a special solution)

$$\Delta T_q = \frac{4l_p^2}{A_H} + \alpha \ln \frac{4l_p^2}{A_H} + O\left(\frac{A_H}{l_p^2}\right) + \text{const.} \tag{3.1}$$

This article points out that when the boundary conditions $\frac{T}{T_c} = z$ (when z is plural) are preset, bosons can produce Bose condensation without an energy layer. Under Bose condensation, the incident wave may condense in the Schwarzschild black hole. At that time, the Schwarzschild black hole event horizon Potential barriers can be generated nearby, and we think that Schwarzschild black holes can also generate superradiation phenomena (This article uses the natural unit system). This implies that the cosmic censorship conjecture may be violated.

3.2 The superradiation effect and uncertainty principle

We find the Klein-Gordon equation [10]

$$\Phi_{;\mu}{}^{;\mu} = 0,\tag{3.2}$$

where we defined $\Phi_{;\mu} \equiv (\partial_\mu - ieA_\mu)\Phi$ and e is the charge of the scalar field. We get $A^\mu = \{A_0(x), 0\}$, and $eA_0(x)$ can be equal to μ (where μ is the mass).

$$A_0 \to \begin{cases} 0 & \text{as } x \to -\infty \\ V & \text{as } x \to +\infty \end{cases}.\tag{3.3}$$

With $\Phi = e^{-i\omega t} f(x)$, which is determined by the ordinary differential equation

$$\frac{d^2 f}{dx^2} + (\omega - eA_0)^2 f = 0.\tag{3.4}$$

We see that particles coming from $-\infty$ and scattering off the potential with reflection and transmission amplitudes $\mathcal{R}$ and $\mathcal{T}$ respectively. With these boundary conditions, the solution to behaves asymptotically as

$$f_{\text{in}}(x) = \mathcal{I}e^{i\omega x} + \mathcal{R}e^{-i\omega x}, x \to -\infty,\tag{3.5}$$

$$f_{\text{in}}(x) = \mathcal{T}e^{ikx}, x \to +\infty\tag{3.6}$$

where $k = \pm(\omega - eV)$.

To define the sign of ω and k we must look at the wave's group velocity. We require $\partial\omega/\partial k > 0$, so that they travel from the left to the right in the x–direction and we take $\omega > 0$.

The reflection coefficient and transmission coefficient depend on the specific shape of the potential A_0. However one can easily show that the Wronskian

$$W = \tilde{f}_1 \frac{d\tilde{f}_2}{dx} - \tilde{f}_2 \frac{d\tilde{f}_1}{dx},\tag{3.7}$$

between two independent solutions, $\tilde{f}_1$ and $\tilde{f}_2$, of is conserved. From the equation on the other hand, if f is a solution then its complex conjugate f^* is another linearly independent solution. We find $|\mathcal{R}|^2 = |\mathcal{I}|^2 - \frac{\omega-eV}{\omega}|\mathcal{T}|^2$. Thus, for $0 < \omega < eV$, it is possible to have superradiant amplification of the reflected current, i.e, $|\mathcal{R}| > |\mathcal{I}|$. There are other potentials that can be completely resolved, which can also show superradiation explicitly.

Classical superradiation effect in the space-time of a steady black hole: we know that $\psi \sim \exp(-i\omega t + im\varphi)$, and the ordinary differential equation

$$\frac{d^2\psi}{dr_*^2} + V\psi = 0.\tag{3.8}$$

We see that particles coming from $-\infty$ and scattering off the potential with reflection and transmission amplitudes $\mathcal{C}$ and $\mathcal{D}$ respectively. With these boundary conditions, the solution to behaves asymptotically as

$$\psi = \begin{cases} Ae^{i\omega_H r_*} + Be^{-i\omega_H r_*}, & r \to r_+ \\ Ce^{i\omega_\infty r_*} + De^{-i\omega_\infty r_*}, & r \to \infty \end{cases}.\tag{3.9}$$

The reflection coefficient and transmission coefficient depend on the specific shape of the potential V. We show that the Wronskian

$$W \equiv \psi \frac{\mathrm{d}\bar{\psi}}{\mathrm{d}r_*} - \bar{\psi} \frac{\mathrm{d}\psi}{\mathrm{d}r_*}, \tag{3.10}$$

$W(r \rightarrow r_+) = 2i\omega_H(|A|^2 - |B|^2), W(r \rightarrow \infty) = 2i\omega_\infty(|C|^2 - |D|^2)$ is conserved. We find $|C|^2 - |D|^2 = \frac{\omega_H}{\omega_\infty}(|A|^2 - |B|^2)$. Thus, for $\omega_H/\omega_\infty < 0$, it is possible to have superradiant amplification of the reflected current, i.e if $|A| = 0$, $|C| > |D|$. There are other potentials that can be completely resolved, which can also show superradiation explicitly.

The principle of joint uncertainty shows that the joint measurement of position and momentum is impossible, that is, the simultaneous measurement of position and momentum can only be an approximate joint measurement, and the error follows the inequality $\Delta x \Delta p \geq 1/2$ (in natural unit system). We find $|\mathcal{R}|^2 = |\mathcal{I}|^2 - \frac{\omega - eV}{\omega}|\mathcal{T}|^2$, and we know that $|\mathcal{R}|^2 \geq -\frac{\omega - eV}{\omega}|\mathcal{T}|^2$ is a necessary condition for the inequality $\Delta x \Delta p \geq 1/2$ to be established. We can pre-set the boundary conditions $eA_0(x) = y\omega$ (which can be $\mu = y\omega$), and we see that when y is relatively large (according to the properties of the boson, y can be very large), $|\mathcal{R}|^2 \geq -\frac{\omega - eV}{\omega}|\mathcal{T}|^2$ may not hold. In the end, we can get $\Delta x \Delta p \geq 1/2$ may not hold. Classical superradiation effect in the space-time of a steady black hole, generalized uncertainty principle may not hold. The same goes for reverse inference.

3.3 Similar to the ADS space and violated uncertainty principle

We can pre-set the boundary conditions $eA_0(x) = -y\omega$ (which can be $\mu = y\omega$). If the boundary conditions of the incident boson are set in advance, the two sides of the probability flow density equation are not equal because of the boundary conditions. Implying a certain probability, this also explains why the no-hair theorem is invalid in quantum effects.

The previous literature [10] indicates that the superradiation effect is a process of entropy subtraction.

Spherical quantum solution in vacuum state.

In this theory, the general relativity theory's field equation is written completely.

$$R_{\mu\nu} - \frac{1}{2}g_{\mu\nu}R = -\frac{8\pi G}{c^4}T_{\mu\nu} \tag{3.11}$$

The Ricci tensor is by $T_{\mu\nu} = 0$ in vacuum state.

$$R_{\mu\nu} = 0 \tag{3.12}$$

The proper time of spherical coordinates is

$$\mathrm{d}\tau^2 = A(t,r)\mathrm{d}t^2 - \frac{1}{c^2}[B(t,r)\mathrm{d}r^2 + r^2\mathrm{d}\theta^2 + r^2\sin\theta\mathrm{d}\varphi^2] \tag{3.13}$$

We obtain the Ricci-tensor equations.

$$R_{tt} = -\frac{A'}{2B} + \frac{A'B'}{4B^2} - \frac{A'}{Br} + \frac{A'^2}{4AB} + \frac{\ddot{B}}{2B} - \frac{\dot{B}^2}{4B^2} - \frac{\dot{A}\dot{B}}{4AB} = 0 \tag{3.14}$$

$$R_{rr} = \frac{A'}{2A} - \frac{A'^2}{4A^2} - \frac{A'B'}{4AB} - \frac{B'}{Br} + \frac{\ddot{B}}{2A} - \frac{\dot{A}\dot{B}}{4A^2} + \frac{\dot{B}^2}{4AB} = 0, \tag{3.15}$$

$$R_{\theta\theta} = -1 + \frac{1}{B} - \frac{rB'}{2B^2} + \frac{rA'}{2AB} = 0, R_{\varphi\varphi} = R_{\theta\theta} \sin^2\theta = 0, R_{tr} = -\frac{\dot{B}}{Br} = 0,$$

$$R_{t\theta} = R_{t\varphi} = R_{r\theta} = R_{r\varphi} = R_{\theta\varphi} = 0 \tag{3.16}$$

In this time, $' = \frac{\partial}{\partial r}, \cdot = \frac{1}{c}\frac{\partial}{\partial t}$,

$$\dot{B} = 0 \tag{3.17}$$

We see that,

$$\frac{R_{tt}}{A} + \frac{R_{rr}}{B} = -\frac{1}{Br}\left(\frac{A'}{A} + \frac{B'}{B}\right) = -\frac{(AB)'}{rAB^2} = 0 \tag{3.18}$$

Hence, we obtain this result.

$$A = \frac{1}{B} \tag{3.19}$$

If,

$$R_{\theta\theta} = -1 + \frac{1}{B} - \frac{rB'}{2B^2} + \frac{rA'}{2AB} = -1 + \left(\frac{r}{B}\right)' = 0 \tag{3.20}$$

If we solve the Eq,

$$\frac{r}{B} = r + C \rightarrow \frac{1}{B} = 1 + \frac{C}{r} \tag{3.21}$$

When r tends to infinity, and we set $C = -ye^y$, Therefore,

$$A = \frac{1}{B} = 1 + \frac{y}{r}\Sigma, \Sigma = e^y \tag{3.22}$$

$$d\tau^2 = \left(1 + \frac{y}{r}\sum\right)dt^2 \tag{3.23}$$

In this time, if particles' mass are m_i, the fusion energy is e,

$$E = Mc^2 = m_1 c^2 + m_2 c^2 + \ldots + m_n c^2 + e \tag{3.24}$$

The effect after the preset boundary is similar to that of Ads cosmological constant. For [11, 12]

$$|\langle \psi_A^\omega \mid \psi_A^* \rangle| \geq 1 - \varepsilon_A^2/2 \tag{3.25}$$

The optimizer is of the form

$$|\psi_A^*\rangle = \frac{(1 - |\alpha|)|\psi_A^\omega\rangle + \alpha A|\psi_A^\omega\rangle}{c_\alpha}, \alpha \in [-1, 1] \tag{3.26}$$

$$|\langle \psi_A^\omega \mid \psi_A^* \rangle| = \frac{1}{c_\alpha}(1 - |\alpha| + \alpha A) \tag{3.27}$$

and

$$|\langle \psi_A^\omega|A|\psi_A^*\rangle| = \frac{1}{c_\alpha}((1 - |\alpha|)A + \alpha\langle A^2\rangle) \tag{3.28}$$

This then proves the assertion.

A general explanation of the uncertainty principle:

$$\sigma_A^2 \sigma_D^2 \geq \left| \frac{1}{2} \langle \{A, D\} \rangle - \langle A \rangle \langle D \rangle \right|^2 + \left| \frac{1}{2i} \langle [A, D] \rangle \right|^2 \tag{3.29}$$

If you multiply the formula by a number less than 1, then the extreme value of the uncertainty principle will become smaller. $|\langle \psi_A^\omega \mid \psi_B^* \rangle|$ do not equal to 0.

The principle of joint uncertainty shows that it is impossible to make joint measurement of position and momentum, that is, to measure position and momentum simultaneously, only approximate joint measurement can be made, and the error follows the inequality $\Delta x \Delta p \geq 1/2$ (in natural unit system). We find $|\mathcal{R}|^2 = |\mathcal{I}|^2 - \frac{\omega - eV}{\omega} |\mathcal{T}|^2$, and we know that $|\mathcal{R}|^2 \geq -\frac{\omega - eV}{\omega} |\mathcal{T}|^2$ is a necessary condition for the inequality $\Delta x \Delta p \geq 1/2$ to be established. We can pre-set the boundary conditions $e A_0(x) = y\omega$ [13] and we see that when y is relatively large (according to the properties of the boson, y can be very large), $|\mathcal{R}|^2 \geq -\frac{\omega - eV}{\omega} |\mathcal{T}|^2$ may not hold. In the end, we can get $\Delta x \Delta p \geq 1/2$ may not hold. When the boundary conditions are assumed and part of the wave function is decoupled, the classical form of the probability function is obtained, and the boundary conditions can be clearly substituted into the uncertainty principle expression, we get a lower uncertainty principle limit. To say the least, because of the limitation of the uncertainty principle, there is a limitation on the value of y, whereas in the literature above, the extreme value of uncertainty principle can be smaller, and the range of the value of y in turn is larger.

3.4   Thermodynamic phase transition

Thermodynamic phase transition. - The state equation of a charged AdS black hole displays a vdW-like thermodynamic behavior. The SBH-LBH coexistence curve has a parametric form [1]

$$\frac{P}{P_c} = \sum_i a_i \left(\frac{T}{T_c} \right)^i. \tag{3.30}$$

The concept of entropy was proposed by the German physicist Clausius in 1865. Kjeldahl defines the increase and decrease of entropy in a thermodynamic system: the total amount of heat used at a constant temperature in a reversible process, and can be expressed as:

$$\Delta S = \frac{\Delta Q}{T} \tag{3.31}$$

if $\frac{T}{T_c} = z$, when z is plural. The Laurent series of the function $f(z)$ with respect to point c is given by:

$$f(z) = \sum_{n=-\infty}^{\infty} a_n (z - c)^n \tag{3.32}$$

It is defined by the following curve integral, which is a generalization of the Cauchy integral formula:

$$a_n = \frac{1}{2\pi i} \oint_\gamma \frac{f(z) dz}{(z - c)^{n+1}} \tag{3.33}$$

Since the algebra of real quaternions is the only fourdimensional division algebra, we introduce the fourdimensional quaternion manifold [2]. When the boundary conditions are preset, with the evolution process, a stable pull equation is formed, and a quaternion time variable can be created for a short time,

$$\tau^4 = (\hat{\tau}_0, \vec{\tau}_1, \vec{\tau}_2, \vec{\tau}_3) = (\hat{\imath}_0\tau_0, \vec{\imath}_1\tau_1, \vec{\imath}_2\tau_2, \vec{\imath}_3\tau_3) \tag{3.34}$$

$$\begin{cases} \hat{\imath}_0\hat{\imath}_0 = \hat{\imath}_0 = 1 \\ \vec{\imath}_1\vec{\imath}_1 = \vec{\imath}_2\vec{\imath}_2 = \vec{\imath}_3\vec{\imath}_3 = \vec{\imath}_1\vec{\imath}_2\vec{\imath}_3 = -\hat{\imath}_0 = -1 \\ \vec{\imath}_1\vec{\imath}_2 = \vec{\imath}_3, \ \vec{\imath}_2\vec{\imath}_3 = \vec{\imath}_1, \ \vec{\imath}_3\vec{\imath}_1 = \vec{\imath}_2, \\ \vec{\imath}_2\vec{\imath}_1 = -\vec{\imath}_3, \ \vec{\imath}_3\vec{\imath}_2 = -\vec{\imath}_1, \vec{\imath}_1\vec{\imath}_3 = -\vec{\imath}_2 \end{cases} \tag{3.35}$$

$$t = \left(\hat{\imath}_0 t_0, \vec{\imath}_1\frac{x_1}{c}, \vec{\imath}_2\frac{x_2}{c}, \vec{\imath}_3\frac{x_3}{c} \right) \tag{3.36}$$

$$\begin{cases} t = t\left(\frac{t_0}{t}, \frac{v}{c}\right) = t(\cos\theta, \vec{\imath}\sin\theta) = t\exp(\vec{\imath}\theta) \\ \bar{t} = t\left(\frac{t_0}{t}, -\frac{v}{c}\right) = t(\cos\theta, -\vec{\imath}\sin\theta) = t\exp(-\vec{\imath}\theta) \end{cases} \tag{3.37}$$

$$\begin{cases} t = \frac{t_0}{\sqrt{1-\frac{v^2}{c^2}}} \exp(\vec{\imath}\theta) \\ \bar{t} = \frac{t_0}{\sqrt{1-\frac{v^2}{c^2}}} \exp(-\vec{\imath}\theta) \end{cases} \tag{3.38}$$

3.5 Schwarzschild black hole produce super-radiation phenomena

According to traditional theory, the Schwarzschild black hole does not produce superradiation. If the boundary conditions are set in advance, the possibility is combined with the wave function of the coupling of the boson in the Schwarzschild black hole, and the mass of the incident boson acts as a mirror, so even if the Schwarzschild black hole can also produce super-radiation phenomena.

Schrödinger's equation for the motion of a particle in a centrally symmetric field is

$$\Delta\psi + (2m/(\hbar)^2)(E - U(r))\psi = 0. \tag{3.39}$$

Let us consider the following radial equation:

$$\frac{1}{r^2}\frac{d}{dr}\left(r^2\frac{dR}{dr}\right) - \frac{l(l+1)}{r^2}R + \frac{2m}{\hbar^2}(E - U(r))R = 0. \tag{3.40}$$

By the substitution

$$R(r) = X(r)/r \tag{3.41}$$

$$\frac{d^2X}{dr^2} + \left[\frac{2m}{\hbar^2}(E - U(r)) - \frac{l(l+1)}{r^2}\right]X = 0. \tag{3.42}$$

For S-wave, $l = 0$, the equation of $X(r)$ is

$$\frac{d^2X}{dr^2} + \frac{2m}{\hbar^2}(E - U(r))X = 0. \tag{3.43}$$

Note that, in the Parikh-Wilczek (solid basic structure on which bigger things can be built), to calculate the self-gravitation reliably the tunnelling particle is carefully thought about/believed as a spherical shell (S-wave). In this way, when it gives off from the black hole the matter-gravity system

33

transits from one spherical state to another. So, the de-Broglie wave function of the emission spherical shell should be

$$\psi(r) = X(r)/r. \tag{3.44}$$

That is, the WKB wave function of a particle can be written as

$$\psi(r) = X(r)/r = \frac{1}{r} \exp\left[\frac{iT}{\hbar}\right], \tag{3.45}$$

where

$$T(r) = T_0(r) + \left(\frac{\hbar}{i}\right) T_1(r) + \left(\frac{\hbar}{i}\right)^2 T_2(r) + \cdots. \tag{3.46}$$

Substituting (3.45) into Schrödinger Equation (3.43) yields

$$T_0 = \pm \int_r^r p_r \mathrm{d}r, \tag{3.47}$$

$$2T_0'T_1' + T_0'' = 0, \tag{3.48}$$

$$2T_0'T_2' + (T_1')^2 + T_1'' = 0, \tag{3.49}$$

where we use a prime to denote differentiation with respect to r.

In region I, we take the WKB wave function as follows

$$\begin{aligned} X_I(r) &= \frac{2}{\sqrt{v}} \sin\left[\frac{1}{\hbar} \int_r^a p_r \mathrm{d}r + \frac{\pi}{4}\right] \\ &= \frac{1}{i\sqrt{v}} \left\{\exp\left[\frac{i}{\hbar} \int_r^a p_r \mathrm{d}r + \frac{i\pi}{4}\right] - \exp\left[-\frac{i}{\hbar} \int_r^a p_r \mathrm{d}r - \frac{i\pi}{4}\right]\right\}, \end{aligned} \tag{3.50}$$

where v is the speed of the tunnelling particle. In area II, the WKB wave function is a linear combination of real (increasing more and more as time goes on) s. (thinking about/when one thinks about) the connection between the swinging back and forth and the (increasing more and more as time goes on) solutions at $r = a$, the WKB wave function in area II can be written as

$$X_{II}(r) = \frac{1}{\sqrt{v}} \exp\left[-\frac{1}{\hbar}\left|\int_a^b p_r \mathrm{d}r\right|\right] \exp\left[-\frac{1}{\hbar}\left|\int_b^r p_r \mathrm{d}r\right|\right]. \tag{3.51}$$

And the WKB wave function in region III is

$$X_{III}(r) = -\frac{1}{\sqrt{v}} \exp\left[-\frac{1}{\hbar}\left|\int_a^b p_r \mathrm{d}r\right|\right] \exp\left[\frac{i}{\hbar} \int_b^r p_r \mathrm{d}r + \frac{i\pi}{4}\right]. \tag{3.52}$$

The probability of barrier penetration is

$$\Gamma_p = \frac{j_{\text{out}}}{j_{\text{in}}} = \frac{v|\psi_{\text{out}}|^2}{v|\psi_{\text{in}}|^2} = \frac{v(X_{\text{out}}(b)/b)^2}{v(X_{\text{in}}(a)/a)^2} = \frac{a^2}{b^2} \cdot \exp\left[-\frac{2\,\text{Im}\,T_0}{\hbar}\right]. \tag{3.53}$$

Let's now calculate the phase space factor corresponding to the black hole tunnelling. For Schwarzschild black hole, the line element in Painlevé coordinates is

$$\mathrm{d}s^2 = -c^2\left(1 - \frac{2MG}{c^2 r}\right)\mathrm{d}t^2 + 2c\sqrt{\frac{2MG}{c^2 r}}\mathrm{d}t\mathrm{d}r + \mathrm{d}r^2 + r^2(\mathrm{d}\theta^2 + \sin^2\theta\mathrm{d}\varphi^2), \tag{3.54}$$

34

and the radial null geodesics are

$$\dot{r} = \frac{dr}{dt} = \pm c\left(1 - \sqrt{\frac{2MG}{c^2 r}}\right).$$ (3.55)

The massive quanta doesn't follow radial-lightlike geodesics (3.55). We treat the massive particle as a de Broglie wave and obtain the expression of $\dot{r}$. Namely,

$$\dot{r} = v_p = \frac{1}{2}v_g = -\frac{1}{2}\frac{g_{00}}{g_{01}} = \frac{1}{2r}\frac{c^2 r^2 - 2MGr}{\sqrt{2MGr}}.$$ (3.56)

Note that to calculate the emission rate correctly, we should take into account the self-gravitation of the tunnelling particle with energy ω.

The canonical momentum p_r and the imaginary part of the action $\mathrm{Im}\, T_0$ can be easily obtained. Namely,

$$p_r = \int_0^{p_r} dp_r' = \int \frac{dH}{\dot{r}} = -i\pi \frac{\hbar}{l_p^2} r,$$ (3.57)

$$\mathrm{Im}\, T_0 = \int_{r_i}^{r_f} p_r dr = -\frac{1}{2}\hbar\left[\frac{4l_p^2}{A_f} - \frac{4l_p^2}{A_i}\right].$$ (3.58)

where $l_p^2 = \frac{\hbar G}{c^3}$.

$$\Gamma(i \to f) = \Gamma_v \cdot \Gamma_p = \Gamma_v \cdot \exp\left[\left(\frac{4l_p^2}{A_f} - \ln\frac{4l_p^2}{A_f}\right) - \left(\frac{4l_p^2}{A_i} - \ln\frac{4l_p^2}{A_i}\right)\right].$$ (3.59)

Thus, we obtain

$$\text{phase space factor} = \exp\left[\left(\frac{4l_p^2}{A_f} - \ln\frac{4l_p^2}{A_f}\right) - \left(\frac{4l_p^2}{A_i} - \ln\frac{4l_p^2}{A_i}\right)\right].$$ (3.60)

If we bear in mind that

$$\text{phase space factor} = \frac{N_f}{N_i} = \frac{e^{T_f}}{e^{T_i}} = e^{T_f - T_i},$$ (3.61)

we naturally get the expression of the black hole temperature to the first order correction (At that time there was a special solution)

$$\Delta T_q = \frac{4l_p^2}{A_H} - \ln\frac{4l_p^2}{A_H}.$$ (3.62)

3.6 second order correction to the black hole entropy

Namely [9],

$$X(r) = \exp\left[\frac{iS_0(r)}{\hbar} + S_1(r) + \frac{\hbar}{i}S_2(r)\right],$$ (3.63)

where

$$S_2 = \int^r -\frac{(S_1'^2 + S_1'')}{2S_0'}dr.$$ (3.64)

35

Like the treatment in Section II, the wave function in region I can be taken as

$$
\begin{aligned}
X_I(r) &= \frac{2}{\sqrt{v}} \sin\left[\frac{1}{\hbar}\left(\int_r^a p_r dr - \hbar^2 S_2(r)\right) + \frac{\pi}{4}\right] \\
&= \frac{1}{i\sqrt{v}}\left\{\exp\left[\frac{i}{\hbar}\left(\int_r^a p_r dr - \hbar^2 S_2(r)\right) + \frac{i\pi}{4}\right] - \exp\left[-\frac{i}{\hbar}\left(\int_r^a p_r dr - \hbar^2 S_2(r)\right) - \frac{i\pi}{4}\right]\right\}.
\end{aligned}
$$

(3.65)

In this region the expression of $S_2(r)$ is

$$
S_2 = \int_r^a -\frac{(S_1'^2 + S_1'')}{2S_0'} dr.
$$

(3.66)

In order to reduce to the first order approximation case, the connexion between the oscillating and the exponential solutions at $r = a$ should be

$$
\frac{2}{\sqrt{v}} \sin\left[\frac{1}{\hbar}\left(\int_r^a p_r dr - \hbar^2 S_2(r)\right) + \frac{\pi}{4}\right] \rightleftharpoons \frac{1}{\sqrt{v}} \exp\left[-\frac{1}{\hbar}\left(\int_a^r |p_r| dr - \hbar^2 S_2(r)\right)\right].
$$

(3.67)

$$r < a \qquad\qquad r > a$$

The expression of $S_2(r)$ is

$$
S_2 = \int_a^r -\frac{(S_1'^2 + S_1'')}{2S_0'} dr.
$$

(3.68)

The connexion at $r = b$ is

$$
\frac{1}{\sqrt{v}} \exp\left[\frac{1}{\hbar}\left(\left|\int_b^r p_r dr\right| - \hbar^2 S_2\right)\right] \rightleftharpoons -\frac{1}{\sqrt{v}} \exp\left[\frac{i}{\hbar}\left(\int_b^r p_r dr - \hbar^2 S_2\right) + \frac{i\pi}{4}\right],
$$

(3.69)

$$r < b \qquad\qquad r > b$$

and the wave function in region III is

$$
X_{III}(r) = -\frac{1}{\sqrt{v}} \exp\left[-\frac{1}{\hbar}(\mathrm{Im}\, S_0 - \hbar^2\, \mathrm{Im}\, S_2)\right] \exp\left[\frac{i}{\hbar}\left(\int_b^r p_r dr - \hbar^2 S_2\right) + \frac{i\pi}{4}\right],
$$

(3.70)

where

$$
\mathrm{Im}\, S_2 = \mathrm{Im} \int_a^b -\frac{(S_1'^2 + S_1'')}{2S_0'} dr.
$$

(3.71)

Since

$$
\psi(r) = X(r)/r,
$$

(3.72)

in region I, the ingoing flux density is

$$
j_{in} = \frac{-i\hbar}{2m}\left(\psi_{in}\frac{\partial}{\partial r}\psi_{in}^* - \psi_{in}^*\frac{\partial}{\partial r}\psi_{in}\right) = v|\psi_{in}^2| = \frac{1}{a^2},
$$

(3.73)

and in region III the outgoing flux density is

$$
j_{out} = \frac{-i\hbar}{2m}\left(\psi_{out}\frac{\partial}{\partial r}\psi_{out}^* - \psi_{out}^*\frac{\partial}{\partial r}\psi_{out}\right) = v|\psi_{out}^2| = \frac{1}{b^2}\exp\left[-\frac{2}{\hbar}(\mathrm{Im}\, S_0 - \hbar^2\, \mathrm{Im}\, S_2)\right].
$$

(3.74)

36

Therefore,

$$\Gamma_p = j_{\text{out}}/j_{\text{in}} = \frac{a^2}{b^2} \exp\left[-\frac{2}{\hbar}(\text{Im } S_0 - \hbar^2 \text{Im } S_2)\right]. \tag{3.75}$$

For Schwarzschild black hole tunnelling, in classically inaccessible region, we have

$$S_0' = p_r = -i\pi\frac{\hbar}{l_p^2}r, \quad S_0'' = -i\pi\frac{\hbar}{l_p^2}, \tag{3.76}$$

and

$$S_1' = -\frac{1}{2}\frac{S_0''}{S_0'} = -\frac{1}{2r}, \quad S_1'' = \frac{1}{2r^2}. \tag{3.77}$$

From (3.77) we can easily obtain

$$S_2' = -\frac{1}{2S_0'}(S_1'^2 + S_1'') = -\left(\frac{3i}{8\pi}\frac{l_p^2}{\hbar}\right) \cdot \frac{1}{r^3}. \tag{3.78}$$

So,

$$S_2 = \int_{r_i}^{r_f} S_2' dr = \frac{3i}{4\hbar}\left(\frac{l_p^2}{A_f} - \frac{l_p^2}{A_i}\right). \tag{3.79}$$

$$\text{phase space factor} = \exp\left[\left(\frac{A_f}{4l_p^2} - \ln\frac{A_f}{4l_p^2} + \frac{3}{2}\frac{l_p^2}{A_f}\right) - \left(\frac{A_i}{4l_p^2} - \ln\frac{A_i}{4l_p^2} + \frac{3}{2}\frac{l_p^2}{A_i}\right)\right]. \tag{3.80}$$

$$\Delta T_q = \frac{4l_p^2}{A_H} - \ln\frac{4l_p^2}{A_H} + \frac{3}{2}\frac{A_H}{l_p^2} + \text{const.} \tag{3.81}$$

We see that under the background of the Schwarzschild black hole, its thermodynamic temperature evolves into a constant. At that time, there is a potential barrier near the horizon. We know that the Schwarzschild black hole can be superradiation at that time (At that time there was a special solution).

An exact solution to the scalar-Einstein equations $R_{ab} = 2\varphi_{A(a)}\varphi_{B(b)}$ which forms a counterexample to many formulations of the cosmic censorship hypothesis was found by Mark D. Roberts in [14]

$$ds^2 = -(1+2\sigma)dv^2 + 2dvdr + r(r - 2\sigma v)(d\theta^2 + \sin^2\theta d\varphi^2), \phi = \frac{1}{2}\ln\left(1 - \frac{2\sigma v}{r}\right). \tag{3.82}$$

where σ is a constant.

In addition, if the naked singularity does not exist, then the universe will become deterministic-it is possible to depend only on the state of the universe at a certain moment (more precisely, a space-like three-dimensional hypersurface called the Cauchy surface State), inferring the entire evolution of the universe (perhaps it is necessary to exclude the limited space hidden in the horizon of the singular point). If the cosmological examination hypothesis fails, it will lead to the definite failure of the universe, because it is impossible to derive the space-time behavior of the universe from the causality of the singular point. The cosmic censorship hypothesis is a formal concern of the physics community. When referring to the event horizon of a black hole, some form of cosmological censorship hypothesis is always involved. A recent article [15] showed that the entropy of a black hole can be reduced under superradiation. This implies that the cosmic censorship conjecture may be violated.

3.7 Summary

In this paper, when the preset boundary condition-the ratio of the temperature of the two systems is a complex number, the entropy can be constructed into the ring structure of an algebraic system, and the entropy is in the same direction as the time dimension. Then we see that under the background of the Schwarzschild black hole, its thermodynamic temperature changes (and gets better) into a constant. At that time, there is a potential barrier near the horizon. We know that the Schwarzschild black hole can be superradiation at that time. This implies that the cosmic censorship conjecture may be violated.

3.8 Acknowledgements

I would like to thank Jing-Yi Zhang. This work is partially supported by National Natural Science Foundation of China (No. 11873025).

Bibliography

[1] Wei, Shao-Wen, and Yu-Xiao Liu. "Clapeyron equations and fitting formula of the coexistence curve in the extended phase space of charged AdS black holes." Physical Review D 91.4 (2015): 044018.

[2] Ariel, Viktor. "Quaternion Space-Time and Matter." arXiv preprint arXiv: 2106.06394 (2021).

[3] Murata K, Soda J. Hawking Radiation from Rotating Black Holes and Gravitational Anomalies[J]. Physical Review D, 2006, 74(4): 200-206.

[4] Medved, A. J. M., and Elias C. Vagenas. "On Hawking radiation as tunneling with logarithmic corrections." Modern Physics Letters A 20.23 (2005): 1723-1728.

[5] M. K. Parikh, F. Wilczek, *Phys. Rev. Lett.*, 85, 5042(2000) [arxiv: hep-th/9907001].

[6] M. K. Parikh, *Int. J. Mod. Phys.* D 13, 2355(2004) [arXiv: hep-th/0405160].

[7] M. K. Parikh, arXiv: hep-th/0402166.

[8] S. Hemming, E. Keski-Vakkuri, *Phys. Rev. D64*, 044006(2001).

[9] Zhang, Jingyi. "Black hole quantum tunnelling and black hole entropy correction." Physics Letters B 668.5 (2008): 353-356.

[10] Brito, Richard, Vitor Cardoso, and Paolo Pani. Superradiance. Springer International Publishing, 2020.

[11] Zoufal, Christa et al. "Error Bounds for Variational Quantum Time Evolution." (2021).

[12] Chen, Wen-Xiang. "Uncertainty principle and super-radiance." Available at SSRN 3683008 (2020).

[13] Chen, Wen-Xiang, and Zi-Yang Huang. "Superradiant stability of the kerr black hole." International Journal of Modern Physics D 29.01 (2020): 2050009.

[14] Roberts, M. D. (1989). "Scalar field counterexamples to the cosmic censorship hypothesis". General Relativity and Gravitation. Springer Science and Business Media LLC. 21 (9): 907-939. Bibcode:1989GReGr..21..907R. doi:10.1007/bf00769864. ISSN 0001-7701. S2CID 121601921.

[15] Chen, Wen-Xiang. "The possibility of the no-hair theorem being violated." Available at SSRN 3569639 (2020).

Chapter 4

The cooperation between loop quantum gravity and Entropic gravity

Abstract: In this paper, we show that bosons can produce bochromatic condensation without an energy layer when the boundary condition $\frac{T}{T_c} = z$ (when z is complex) is preset. $r_R = z r_A \left(1 - \frac{v^2}{c^2}\right)^{\frac{1}{2}}$ (when z is A complex number). A new form of gravitational potential is obtained by combining the theory of gravitational entropy with loop quantum gravity.

Keywords: quaternion, loop quantum gravity, Entropic gravity

4.1 Introduction

The so-called new interpretation of the entropy of gravity belongs to a non-mainstream branch of gravity research for a long time. The core idea is that gravity is a statistical phenomenon; it is not fundamental, like the other three basic interactions. The statistical origin of gravity first came from the hint of black hole thermodynamics. For example, the famous "Bekenstein-Hawking" entropy formula in the early 1970s. The more famous work followed was that in the mid-1990s, T. Jacobson derived Einstein's equations based on the thermodynamics of black holes-the basic equations of general relativity. That is, Einstein's equation is actually the state equation of the space-time thermodynamic system. The most recent development is the entropy interpretation of gravity mentioned by the subject, which was proposed by E. Verlinde in early 2010. At that time, it attracted considerable attention. "Entropy force" is a macroscopic force that describes the system's response to changes in entropy. A typical example is the elasticity of rubber bands. The entropy of Verlinde can be roughly explained as: there is a (hypothetical) holographic surface between two objects. When the distance between the two objects changes, the entropy on this holographic surface will also change, resulting in a change in system energy. The system always tends to increase in entropy, so gravity is reflected in the response to this change. In this explanation, the holographic principle is the basic principle, while the general theory of relativity is presented layer by layer. Entropy belongs to the so-called "emergent phenomena" in condensed matter physics, that is, a phenomenon that only appears on a certain level.

General relativity and quantum theory have profoundly changed our view of the world. Moreover, in the past few decades, these two theories have been accurately confirmed. Loop quantum gravity takes this novel viewpoint seriously: that is, the concept of time and space in general relativity is directly incorporated into quantum field theory, and the result is a theory that is fundamentally different from traditional quantum field theory. It not only provides an accurate mathematical picture of quantum space-time, but also answers some long-standing questions, such as the thermodynamics of black holes and the physics of the big bang.

The most attractive aspect of loop quantum gravity is that it predicts that space is not infinitely divisible, but has a granular structure. In this theory, the size of these basic "space quanta" can be clearly calculated, just like the energy level structure of a hydrogen atom. In the past 50 years or so, people have explored many ways to construct a quantum theory of gravity, but only two have given a complete mathematical description of the quantum properties of gravitational fields: loop quantum gravity and string theory. In the last decade, there have been major advances in loop gravity and string theory, but it is worth emphasizing that both theories imply unsolved problems. More importantly, none of them has been confirmed by experiments. There is hope that direct experimental support will come soon, but any current theory may be right, partly right or simply wrong. In any case, the fact that we have two complete, temporary theories of quantum gravity is encouraging. We are neither in the dark nor lost in a large number of alternative theories. Quantum gravity provides a window into the basic structure of nature.

Generalized covariance – the laws of physics can be expressed in any coordinate system, and this is the basic assumption of general relativity.

Background independence - There is no independent invariant metric, coordinate system, etc. that can be used as a background.

Loop quantum gravity also assumes that the basic principles of quantum theory are correct. Generalized covariant theory including the general theory of relativity, for example, the generalized covariant theory including the special theory of relativity (special covariance), the background independent theory including Newtonian mechanics (assuming the same timeline) independently, special theory of relativity (its background is minkowski space, background metric is minkowski metric), electronic movement in the background of electromagnetic field equations, etc., Background independent theories include general relativity, in which the value of the metric tensor is entirely determined.

In the reference frame of a specific superradiation (Loop quantum gravity theory), in its self-reference frame, γ modifies the structure. In the free space where $\gamma < 1$, the electron is not a stable particle, and its wave function diverges and grows. However, in the binding state, this divergence is controlled by the binding potential, and γ can take a value less than 1.

If the boundary conditions are preset, the boundary conditions $r_R = z r_A \left(1 - \frac{v^2}{c^2}\right)^{\frac{1}{2}}$ (when z is plural), the effective action form satisfies the effective action form of Hawking radiation, and is not necessarily at the boundary of event horizon. We combine the entropy hypothesis of gravity with loop quantum gravity to obtain a preliminary new form of gravitational potential:

$$\Phi(r) = -\frac{iGm^2}{r}\left(\frac{1}{K}(e^{\frac{i2Gm}{rc^2}} - 1) + 1\right)$$

(4.1)

4.2 Thermodynamic phase transition

Between these two points A and R, there are two possible clock measurements, which can be simply calculated as follows: [3]

1. Relative gravitational redshift, expressed by the ratio of two different points

$$\frac{z_A}{z_R} = \frac{\left(1 - \frac{r_s}{r_A}\right)^{-1/2} - 1}{\left(1 - \frac{r_s}{r_R}\right)^{-1/2} - 1} \tag{4.2}$$

2. The difference in gravitational red-shift at two different points.

$$\Delta z = z_A - z_R = \left(1 - \frac{r_s}{r_A}\right)^{-1/2} - \left(1 - \frac{r_s}{r_R}\right)^{-1/2} \tag{4.3}$$

Thermodynamic phase transition. - The state equation of a charged AdS black hole displays a vdW-like thermodynamic behavior. The SBH-LBH coexistence curve has a parametric form [1]

$$\frac{r_R}{r_A} = \frac{P}{P_c} = \sum_i a_i \left(\frac{T}{T_c}\right)^i. \tag{4.4}$$

The concept of entropy was proposed by the German physicist Clausius in 1865. Kjeldahl defines the increase and decrease of entropy in a thermodynamic system: the total amount of heat used at a constant temperature in a reversible process, and can be expressed as:

$$\Delta S = \frac{\Delta Q}{T} \tag{4.5}$$

if $\frac{T}{T_c} = z$, when z is plural. The Laurent series of the function $f(z)$ with respect to point c is given by:

$$f(z) = \sum_{n=-\infty}^{\infty} a_n (z - c)^n \tag{4.6}$$

It is defined by the following curve integral, which is a generalization of the Cauchy integral formula:

$$a_n = \frac{1}{2\pi i} \oint_\gamma \frac{f(z)\mathrm{d}z}{(z - c)^{n+1}} \tag{4.7}$$

Since the algebra of real quaternions is the only fourdimensional division algebra, we introduce the fourdimensional quaternion manifold, [2]

$$\tau^4 = (\hat{\tau}_0, \vec{\tau}_1, \vec{\tau}_2, \vec{\tau}_3) = (\hat{\imath}_0 \tau_0, \vec{\imath}_1 \tau_1, \vec{\imath}_2 \tau_2, \vec{\imath}_3 \tau_3) \tag{4.8}$$

$$\begin{cases} \hat{\imath}_0 \hat{\imath}_0 = \hat{\imath}_0 = 1 \\ \vec{\imath}_1 \vec{\imath}_1 = \vec{\imath}_2 \vec{\imath}_2 = \vec{\imath}_3 \vec{\imath}_3 = \vec{\imath}_1 \vec{\imath}_2 \vec{v}_3 = -\hat{\imath}_0 = -1 \\ \vec{\imath}_1 \vec{\imath}_2 = \vec{\imath}_3, \ \vec{\imath}_2 \vec{\imath}_3 = \vec{\imath}_1, \ \vec{\imath}_3 \vec{\imath}_1 = \vec{\imath}_2, \\ \vec{\imath}_2 \vec{\imath}_1 = -\vec{\imath}_3, \ \vec{\imath}_3 \vec{\imath}_2 = -\vec{\imath}_1, \vec{\imath}_1 \vec{\imath}_3 = -\vec{\imath}_2 \end{cases} \tag{4.9}$$

$$\boldsymbol{t} = \left(\hat{\imath}_0 t_0, \vec{\imath}_1 \frac{x_1}{c}, \vec{\imath}_2 \frac{x_2}{c}, \vec{\imath}_3 \frac{x_3}{c}\right) \tag{4.10}$$

$$\begin{cases} t = t\left(\frac{t_0}{t}, \frac{\vec{v}}{c}\right) = t(\cos\theta, \vec{i}\sin\theta) = t\exp(\vec{i}\theta) \\ \bar{t} = t\left(\frac{t_0}{t}, -\frac{\vec{v}}{c}\right) = t(\cos\theta, -\vec{i}\sin\theta) = t\exp(-\vec{i}\theta) \end{cases} \tag{4.11}$$

$$\begin{cases} t = \frac{t_0}{\sqrt{1-\frac{v^2}{c^2}}}\exp(\vec{i}\theta) \\ \bar{t} = \frac{t_0}{\sqrt{1-\frac{v^2}{c^2}}}\exp(-\vec{i}\theta) \end{cases} \tag{4.12}$$

4.3 New INITIAL EQUATIONS

We consider a static spherically symmetric space-time with metric

$$ds^2 = e^z dt^2 - e^\beta dr^2 - r^2[d\theta^2 + \sin^2\theta d\phi^2], \tag{4.13}$$

when z is plural. The gravitational field interacts with real SF $\phi(r)$ described by Lagrangian density

$$L = \frac{1}{2}\partial_\mu\phi\partial^\mu\phi - V(\phi) \tag{4.14}$$

$$\rho = \frac{1}{K}\left(e^{\frac{i2GM}{c^2 r}} - 1\right) \tag{4.15}$$

The density has also been expressed as $\rho = \frac{x_0 - x}{x}$ in terms of the distance measurement x_0 in a region devoid of gravity and x the measurement made after considering the influence of gravitational spatial compression. This leads to

$$x = \frac{x_0}{\frac{1}{K}(e^{\frac{i2Gm}{rc^2}} - 1) + 1} \tag{4.16}$$

The same expression yields gravitational red shift if distance is replaced by wavelength of light λ and is written as

$$\lambda = \frac{\lambda_0}{\frac{1}{K}\left(e^{\frac{i2Gm}{rc^2}} - 1\right) + 1} \tag{4.17}$$

These concepts have been able to reproduce the gravitational red shift and gravitational lensing in the same form as predicted by general relativity.

The Newtonian potential φ is

$$\Phi(r) = -\frac{Gm^2}{r} \tag{4.18}$$

where r is the distance between two masses. On considering the spatial compression, we can replace r by by compressed r under the influence of two gravitating masses such that

$$r_c = \frac{r}{\frac{1}{K}(e^{\frac{i2Gm}{rc^2}} - 1) + 1} \tag{4.19}$$

This leads to the modified potential as

$$\Phi(r) = -\frac{iGm^2}{r}\left(\frac{1}{K}\left(e^{\frac{i2Gm}{rc^2}} - 1\right) + 1\right) \tag{4.20}$$

44

4.4 Summary

This article points out that when the boundary conditions $\frac{T}{T_c} = z$ (when z is plural) are preset, bosons can produce Bose condensation without an energy layer. If the boundary conditions are preset, the boundary conditions $r_R = z r_A \left(1 - \frac{v^2}{c^2}\right)^{\frac{1}{2}}$ (when z is plural), the effective action form satisfies the effective action form of Hawking radiation, and is not necessarily at the boundary of event horizon. We combine the entropy hypothesis of gravity with loop quantum gravity to obtain a preliminary new form of gravitational potential.

Acknowledgements

I would like to thank Jing-Yi Zhang. This work is partially supported by National Natural Science Foundation of China (No. 11873025).

Bibliography

[1] Wei, Shao-Wen, and Yu-Xiao Liu. "Clapeyron equations and fitting formula of the coexistence curve in the extended phase space of charged AdS black holes." Physical Review D 91.4 (2015): 044018.

[2] Ariel, Viktor. "Quaternion Space-Time and Matter." arXiv preprint arXiv: 2106.06394 (2021).

[3] Mir Hameeda, M. C. Rocca. "Gupta-Feynman Based Quantum Theory of Gravity and the Compressed Space". vixra.org:2109.0023.

Chapter 5

Preset boundary conditions and the possibility of making time crystals

Abstract: Using the quaternion algebraic tools widely used at the end of the 19th century, we deduce a novel theory of space-time unity that can enhance the theories of special relativity and general relativity. When the preset boundary condition-the ratio of the temperature of the two systems is a complex number, then the entropy can be constructed into the ring structure of the algebraic system, and the entropy and the time dimension are in the same direction, then it is possible to construct a time crystal.

Keywords: quaternion, kerr blackhole, time crystal

5.1 Introduction

The time crystal is an open system that maintains an unbalanced state with the surrounding environment, showing the characteristics of breaking the symmetry of time translation. A scientific report in March 2017 pointed out that this theoretical concept has been experimentally confirmed; as time goes by, the time crystal still cannot reach thermal equilibrium with the environment.

The concept of time crystals was first proposed by Nobel Prize winner Frank Wilseck in 2012. Compared with ordinary crystals that repeat periodically in space, time crystals repeat periodically in time, showing the state of perpetual motion machines. The time crystal has a spontaneous symmetry breaking phenomenon in the time translation symmetry. Time crystals are also related to zero-point energy and dynamic Casimir effects.

In 2016, the Department of Physics, University of California, Berkeley, Yao Ying (English: Norman Y. Yao) and colleagues proposed a blueprint for building time crystals in the laboratory; this blueprint was subsequently adopted by two groups including Christopher Monroe and the University of Maryland. Mikhail Lukin of Harvard University, both teams successfully created time crystals. The results of the experiment were published in the journal Nature in March 2017.

Conventional crystals are three-dimensional objects in which atoms are arranged repeatedly in a regular manner. Time crystals are crystals with more than four dimensions and have a periodic structure of time and space. Time crystals can spontaneously destroy the symmetry of time translation and perform non-translational movement in space. The composition of time crystals is composed of non-localized particles and interrelated motions in "space". It is the "extra dimension" of energy-saving particles that transcends "fixedness". The energy and momentum of the domain space and the existence of time crystals also reveal the meaning of the existence of "super extra dimensions".

It will change over time, but it will continue to return to its original shape, just as the hands of a clock periodically return to their original positions. Unlike ordinary clocks or other periodic processes, time crystals and space crystals have the least energy. You can think of it as a clock that can keep the time accurate forever, even after the universe heats to death.

We switched from quaternion algebra to quaternion calculus by introducing an appropriate form of quaternion gradient operator. We apply this differentiation process to the general quaternion potential to derive the unified form of the force field and Maxwell's equations. New components in the interaction of electromagnetic and gravitational forces may lead to exciting new discoveries of physical phenomena.

Therefore, using the quaternion algebraic tools widely used at the end of the 19th century, we deduce a novel theory of space-time unity that can enhance the theories of special relativity and general relativity.

When the preset boundary condition-the ratio of the temperature of the two systems is a complex number, then the entropy can be constructed into the ring structure of the algebraic system, and the entropy and the time dimension are in the same direction, then it is possible to construct a time crystal.

5.2 The possibility of time crystals

We make a hypothesis: Thermodynamic phase transition. - The state equation of a charged AdS black hole displays a vdW-like thermodynamic behavior. The SBH-LBH coexistence curve has a parametric form [1]

$$\frac{P}{P_c} = \sum_i a_i \left(\frac{T}{T_c}\right)^i. \tag{5.1}$$

The concept of entropy was proposed by the German physicist Clausius in 1865. Kjeldahl defines the increase and decrease of entropy in a thermodynamic system: the total amount of heat used at a constant temperature in a reversible process, and can be expressed as:

$$\Delta S = \frac{\Delta Q}{T} \tag{5.2}$$

if $\frac{T}{T_c} = z$, when z is plural. The Laurent series of the function $f(z)$ with respect to point c is given by:

$$f(z) = \sum_{n=-\infty}^{\infty} a_n (z - c)^n \tag{5.3}$$

It is defined by the following curve integral, which is a generalization of the Cauchy integral formula:

$$a_n = \frac{1}{2\pi i} \oint_\gamma \frac{f(z)\mathrm{d}z}{(z - c)^{n+1}} \tag{5.4}$$

Since the algebra of real quaternions is the only fourdimensional division algebra, we introduce the fourdimensional quaternion manifold, [2]

$$\tau^4 = (\hat{\tau}_0, \vec{\tau}_1, \vec{\tau}_2, \vec{\tau}_3) = (\hat{\imath}_0 \tau_0, \vec{\imath}_1 \tau_1, \vec{\imath}_2 \tau_2, \vec{\imath}_3 \tau_3) \tag{5.5}$$

$$\begin{cases} \hat{\imath}_0\hat{\imath}_0 = \hat{\imath}_0 = 1 \\ \vec{\imath}_1\vec{\imath}_1 = \vec{\imath}_2\vec{\imath}_2 = \vec{\imath}_3\vec{\imath}_3 = \vec{\imath}_1\vec{\imath}_2\vec{v}_3 = -\hat{\imath}_0 = -1 \\ \vec{\imath}_1\vec{\imath}_2 = \vec{\imath}_3,\ \vec{\imath}_2\vec{\imath}_3 = \vec{\imath}_1,\ \vec{\imath}_3\vec{\imath}_1 = \vec{\imath}_2, \\ \vec{\imath}_2\vec{\imath}_1 = -\vec{\imath}_3,\ \vec{\imath}_3\vec{\imath}_2 = -\vec{\imath}_1,\ \vec{\imath}_1\vec{\imath}_3 = -\vec{\imath}_2 \end{cases} \tag{5.6}$$

$$t = \left(\hat{\imath}_0 t_0, \vec{\imath}_1 \frac{x_1}{c}, \vec{\imath}_2 \frac{x_2}{c}, \vec{\imath}_3 \frac{x_3}{c} \right) \tag{5.7}$$

$$\begin{cases} t = t\left(\frac{t_0}{t}, \frac{\vec{v}}{c}\right) = t(\cos\theta, \vec{\imath}\sin\theta) = t\exp(\vec{\imath}\theta) \\ \bar{t} = t\left(\frac{t_0}{t}, -\frac{\vec{v}}{c}\right) = t(\cos\theta, -\vec{\imath}\sin\theta) = t\exp(-\vec{\imath}\theta) \end{cases} \tag{5.8}$$

$$\begin{cases} t = \frac{t_0}{\sqrt{1-\frac{v^2}{c^2}}} \exp(\vec{\imath}\theta) \\ \bar{t} = \frac{t_0}{\sqrt{1-\frac{v^2}{c^2}}} \exp(-\vec{\imath}\theta) \end{cases} \tag{5.9}$$

5.3 Time-crystal, action path to Kerr Black Hole

We will prove that the Hawking radiation of the Kerr black hole can be understood as the flux that offsets the gravitational anomaly. The key is that near the horizon, the scalar field theory in the spacetime of a 4-dimensional Kerr black hole can be simplified to a 2-dimensional field theory. Since space-time is not spherically symmetric, this is an unexpected result.

In Boyer-Linquist coordinates, Kerr metric reads [3]

$$ds^2 = -\frac{\Delta - a^2\sin^2\theta}{\Sigma}dt^2 - 2a\sin^2\theta\frac{r^2+a^2-\Delta}{\Sigma}dtd\varphi \tag{5.10}$$

$$+ \frac{(r^2+a^2)^2 - \Delta a^2\sin^2\theta}{\Sigma}\sin^2\theta d\varphi^2 + \frac{\Sigma}{\Delta}dr^2 + \Sigma d\theta^2 \tag{5.11}$$

$$\Sigma = r^2 + a^2\cos^2\theta,\ \Delta = r^2 - 2Mr + a^2 = (r-r_+)(r-r_-). \tag{5.12}$$

The action for the scalar field in the Kerr spacetime is

$$\begin{aligned} S[\phi] &= \frac{1}{2}\int d^4x\sqrt{-g}\phi\nabla^2\phi \\ &= \frac{1}{2}\int d^4x\sqrt{-g}\phi\frac{1}{\Sigma}\left[-\left(\frac{(r^2+a^2)^2}{\Delta} - a^2\sin^2\theta\right)\partial_t^2 - \frac{2a(r^2+a^2-\Delta)}{\Delta}\partial_t\partial_\varphi \right. \\ &\quad \left. + \left(\frac{1}{\sin^2\theta} - \frac{a^2}{\Delta}\right)\partial_\varphi^2 + \partial_r\Delta\partial_r + \frac{1}{\sin\theta}\partial_\theta\sin\theta\partial_\theta\right]\phi \end{aligned} \tag{5.13}$$

Taking the limit $r \to r+$ and leaving the dominant terms, we have

$$S[\phi] = \frac{1}{2}\int d^4x\sin\theta\phi\left[-\frac{(r_+^2+a^2)^2}{\Delta}\partial_t^2 - \frac{2a(r_+^2+a^2)}{\Delta}\partial_t\partial_\varphi - \frac{a^2}{\Delta}\partial_\varphi^2 + \partial_r\Delta\partial_r\right]\phi \tag{5.14}$$

Now we transform the coordinates to the locally non-rotating coordinate system by

$$\begin{cases} \psi = \varphi - \Omega_H t \\ \xi = t \end{cases} \tag{5.15}$$

49

where

$$\Omega_H \equiv \frac{a}{r_+^2 + a^2}.$$ (5.16)

We can rewrite the action

$$S[\phi] = \frac{a}{2\Omega_H} \int d^4x \sin\theta\phi \left(-\frac{1}{f(r)}\partial_\xi^2 + \partial_r f(r)\partial_r\right)\phi$$ (5.17)

We know that when $\sin\theta = 0$, the pull equation for action can conform to the above form, but the boundary becomes 0. However, if the boundary conditions are preset, the boundary conditions $\frac{T}{T_c} = z$ act as $\sin\theta$, The effective action form satisfies the effective action form of Hawking radiation, and is not necessarily at the boundary of event horizon. When the boundary condition is preset, a new path is obtained

$$S[z,\phi] = \frac{a}{2\Omega_H} \int d^4xze^z\phi \left(-\frac{1}{f(r)}\partial_\xi^2 + \partial_r f(r)\partial_r\right)\phi.$$ (5.18)

5.4   Summary

In this paper, when the preset boundary condition-the ratio of the temperature of the two systems is a complex number, the entropy can be constructed into the ring structure of an algebraic system, and the entropy is in the same direction as the time dimension, then a time crystal can be constructed.

Acknowledgements

I would like to thank Jing-Yi Zhang. This work is partially supported by National Natural Science Foundation of China (No. 11873025).

Bibliography

[1] Wei, Shao-Wen, and Yu-Xiao Liu. "Clapeyron equations and fitting formula of the coexistence curve in the extended phase space of charged AdS black holes." Physical Review D 91.4 (2015): 044018.

[2] Ariel, Viktor. "Quaternion Space-Time and Matter." arXiv preprint arXiv: 2106.06394 (2021).

[3] Murata K, Soda J. Hawking Radiation from Rotating Black Holes and Gravitational Anomalies[J]. Physical Review D, 2006, 74(4): 200-206.

Publisher: Eliva Press SRL

Email: info@elivapress.com

www.ingramcontent.com/pod-product-compliance
Lightning Source LLC
Chambersburg PA
CBHW051248170526
45165CB00004B/1625